DC Error Analysis of a Discrete Instrumentation Amplifier

Derivations of a Traditional 3-Opamp Topology

DC Error Analysis of a Discrete Instrumentation Amplifier

Derivations of a Traditional 3-Opamp Topology

T. D. McGlone, Ph.D.

tdm@ieee.org

2016

All rights reserved. This book or any portion thereof may not be reproduced or used in any manner whatsoever without the express written permission of the publisher except for the use of brief quotations in a book review or scholarly journal.

Copyright © 2016 by T. David McGlone

ISBN: 978-1-5306-4338-7

First Printing: 2016
Printed in U.S.A.

The text font in this book is Garamond 11 pt. Headings are based on Century Gothic.

Dedicated to the inspiration and guidance of Garry Cleveland
Lockheed Space Systems Div., Sunnyvale, CA

to whom I never took the chance to let know how much his influence
and guidance taught me a level of excellence few others
– individuals or companies –
have approached or accepted since.

Special thanks to Maureen Ericksen for proof-reading

Table of Contents

Introduction 1
   Why? Measurement Quantization 4
Chapter 1: The Instrumentation Amplifier 7
Chapter 2: Symbolic Equations 9
   Differential & Common-Mode Gain Error 9
   Network Stage 1 9
      Stage 1 Differential Gain 10
      Stage 1 Common-Mode Gain 13
   Network Stage 2 17
      2nd-Stage Difference Amplifier Gain 17
      2nd-Stage Differential and Common-Mode Gain 20
      Effect of Resistor Mismatch on ACM2 21
   Offset Voltage and Bias Current Error 22
      Offset Voltage Error: ε_{os} 22
      Inverting Bias Current Error: ε_{ibm} 24
      Bias Current Offset Error: ε_{ibp} 25
   Total 2nd-Stage DC Error Model 26
Chapter 3: Total System Error 28
   System Differential Gain 28
   System Common-Mode Gain: 29
   Total DC Offset Error 30
   Common-Mode Rejection Ratio (CMRR) 30
   Resistor Tolerance & Aging 31
      Tolerance 31
      2nd-Stage Network 32
   Aging 33
Chapter 4: Numerical Example 35
   Expression Summary 35
      1st-Stage: 35
      2nd-Stage: 35
   System Analysis 37
      Operating Parameters 37
      Temperature Assumptions 37

Table of Contents - cont.

Chapter 4: cont.
- Resistor Analysis 38
  - Total Resistor Tolerance 38
  - Resistor Parameters 39
  - Aging Calculations: 40
  - Worst-Case 7yr EOL Resistor Tolerance 42
  - Array Resistor Considerations 42
- Component Parameters 43
  - Opamp Nominal (max) Parameters (25 C) 43
  - Resistor EOL (max) Tolerance (7 yr) (but matching ≤ 0.02%) 43
  - Offset Error 43
  - Bias Current Error 43
- Total Worst-Case Nominal DC Error 43
- Gain & Common-Mode Rejection Ratio (CMRR): 44
- Representative Instrumentation Amplifier – AD524 48

Chapter 5: Phase Error 49

Chapter 6: Supply Voltage regulation 53
- Power-Supply Rejection Ratio (PSRR) 53
  - PSRR of AD524 54
  - PSRR of ISL70417 54
- LM117 Adjustable Positive Voltage Regulator 55

Appendix A: Physical Connection 63

Appendix B: Mathematica Snippets 64
- Define Parameters 64
- Define Plots 66

Appendix C: Opamp Fundamentals 67
- Inverting Configuration 67
- Non-Inverting Configuration 68
- Open-Loop Gain A_{OL} 69
- Offset Voltage 71
- Bias Current 72
- Common-Mode Gain 74
- Input Impedance (Resistance) 75
- Output Impedance (Resistance) 77
- Noise 78

References 79

Preface

I was at work one day and one of the newer engineers came and asked my comments about a circuit he was working on. As it happened, he was developing a circuit similar to one I had once developed for operation in extreme environments when the available qualified parts weren't quite right for the project.

He had correctly implemented the basic circuit but the computer simulations weren't providing the answers he expected. Digging into my files, I discovered a worst-case analysis for an essentially identical circuit.

When I showed him my results – from the days when a hand analysis was used to verify a computer analysis rather than a computer analysis being required to verify hand-analysis, his comment was along the lines of: "*Wow, that's more detail than we bother with.*"

But the devil *is* in the details ...

I started to re-write a copy of these notes for him (the originals were over 20 years old, scribbled on, and handwritten – with coffee stains). By the time I got the work cleaned up with explanations, the project had moved beyond "notes" and into something more extensive ... other co-workers expressed interest in this, perhaps others might have interest as well.

So John and Rene, here it is.

Dave
Boise, Idaho

Introduction

For all the insistence on *"reliability"*; for all the cost and time it takes to obtain that reliability, the models upon which simulations are based are often not held to the same standards or scrutiny, even for intended use in extreme environments - the expense of maintaining and verifying models is often *"beyond the scope of our budget"*.

A "worst-case" analysis is just that: under the worst-case extreme conditions, what are the expected performance limits? What are the possible destruction limits? A worst-case analysis is intended to define these limits. This type of analysis may –will– produce results far different than a "nominal" analysis, the primary design analysis which defines circuit operation when everything is ... nominal.

Having been trained on projects which were inaccessible once deployed, I discovered early on that it was far better to understand the limits of operation – and tweak if necessary – before deployment. Hence, the interest in "worst-case" as opposed to "expected".

Worst-case analysis is expensive; so is tweaking a unit which doesn't meet worst-case performance requirements. Most worst-case conditions don't happen, and as development even for "expected" conditions is expensive enough, it can seem minimizing the effort spent on defining a set of criteria which likely won't occur is a great place to trim a budget.

Until "worst-case" does occur - after deployment.

Models are typically not developed for "worst-case" operation. Manufacturers try to do the best job for their customers, but if a customer wishes to use a component near the edges of operation capability, the models themselves need to be verified by the customer for the desired region of operation.

"caveat emptor", *"you get what you pay for"*, and *"just how valuable is this project anyway?"*

Instrumentation amps are often the core element of a measurement system. More complex than an op-amp, this traditional configuration built of 3 opamps allows the development of a true voltage difference amplifier having "infinite" impedance at both input terminals and near-zero output impedance. The gain may be set with a single resistor; the output may be referenced independently of the input – a method of input/output isolation, a method of level-shifting.

A handy little network.

The very high amplifier input impedance in parallel with the terminal resistance of the sensor will approximate the resistance of only the sensor which helps decrease measurement errors. This topology also has the advantage of suppressing "common-mode" voltages: if a sensor is biased to 1V but has a signal range of 0 to 10 mV, the ideal amplifier would multiply only the differential signal (0 to 10 mV) while completely rejecting the common-mode 1V - the maximum output of the amplifier would be (gain × 10 mV) over a linear range regardless of the common-mode levels.

The output is isolated from the input to a degree expressed by the parameter "CMRR": Common-Mode Rejection Ratio – the ratio of (desired) differential gain to (undesired) common-mode gain. The ideal CMRR is infinite (ideal common-mode gain is 0). CMRR is a factor degrading measurement resolution.

The ideal instrumentation amplifier appears as:

$$V_{out} - REF \equiv GAIN \times (V1 - V2) + 0 \times V_{cm}$$

Introduction

Once upon a very long time ago, I derived these equations out of necessity. Textbooks tend to present expressions *"by letting this value go to 1 and that term go to zero, we get this result"*, *"assume this value goes to infinity"*, or even worse ... *"the derivation is left to the student"*. In the working world, I found I needed to derive *"but it's not 1 and not only does that term not go to zero, it turns out to be a significant error source."* And I no longer had the luxury of time to re-derive expressions for circuits of differing topologies.

I thought this information might be outdated – I was recently chastised for applying "outdated techniques" from 25 years ago (on a 2012 project using the same topology and many of the same parts as one from 1986).

Then I needed an almost identical analysis for a recent project ... because the computer models were rarely sufficient for this type of analysis - Monte Carlo runs or not. The simulations measure the model - but has the model itself been verified?

I hope these derivations are useful in some manner. In general: use a commercially available instrumentation amplifier for extreme condition projects unless you can truly explain why it would be necessary to build a discrete version.

In the example used for this discussion, although the final amplifier was impressive, the Analog Device's AD524 out-performed it for the purpose intended - as would many other available instrumentation amplifiers.

It also cost less, is more reliable, and has a smaller footprint.

Why? Measurement Quantization

Quantization of a signal is at least two steps removed from physical "reality".

The transducing detector itself represents a 1st step of remoteness: it produces an approximate electrical analogue of the physical phenomenon to be measured. The output of this 1st-step analogue is then processed – amplification and filtering at a minimum – in preparation for quantization. A 2nd-step of remoteness from "reality": the phenomenon itself isn't quantized, only an analogue of the analogue is applied to the analog-to-digital converter. Each step degrades the signal, adds a bit of noise - each step, including the quantization process - removes a bit of signal integrity.

The quantizing ADC generates digital data now at least three stages removed from reality - the less degradation of signal between the detector and quantizer, the more "accurate" the digital representation. An instrumentation amplifier can be a critical component of the signal chain.

By definition, the standard deviation of a Gaussian function is equivalent to the RMS value of that function. The RMS value of an AC signal has equal power as that of a DC signal of the same magnitude. The peak-to-peak (pp) variation of a random Gaussian signal is equal to $\pm 3\ \sigma$ (with a peak-to-peak magnitude of 6σ) which represents 99.87% of all occurrences. The variation of 3.3σ contains 99.95% of values - not quite sufficient for 1 LSB of a 12-bit converter (99.98%).

One standard for quantization error is that the 3σ value of uncertainty be no more than ½LSB. If an ideal Gaussian noise source has an RMS value of 1 nV/rtHz with a noise bandwidth of 1 kHz, the RMS noise would be 32 nV. This implies 1LSB has a magnitude of 192 nV. The full-scale range (FSR) of the necessary 12-bit ADC would be 393.2 µV.

Such an ADC is hard to find.

Introduction

Amplification is necessary; unfortunately, uncertainty is amplified along with signal – and the process of amplification and other signal processing such as filtering also places additional uncertainty on the signal. The instrumentation amplifier (IA) is often the first electronic component in the signal chain; a very good IA will amplify the signal without adding excessive noise ... noise being loosely defined as "everything except the desired signal".

The total signal to be quantized is the RMS sum of the signal and various noise and uncertainties added during initial signal processing. If the minimum signal to be measured (as opposed to detected) is 10x the noise floor (20 dB), the measurement signal error would be:

$$V_{measure} = \sqrt{V_{sig}^2 + V_{noyz}^2} = \sqrt{1_{sig}^2 + 0.1_{noyz}^2} \rightarrow \varepsilon \approx 0.5\%$$

If the minimum signal to be measured is 100x the noise floor (40 dB), the signal error would be:

$$V_{measure} = \sqrt{V_{sig}^2 + V_{noyz}^2} = \sqrt{1_{sig}^2 + 0.01_{noyz}^2} \rightarrow \varepsilon \approx 0.005\%$$

A reasonable allocation of dynamic range would be a minimum measure signal[1] at 20 dB above the noise floor with signal max to min of 100. This represents a 60 dB quantization range; the number of bits required to represent this range would be:

$$ENOB = \frac{(S_{max,dB} + N_{dB}) - 1.76}{6.02} = \frac{(40 + 20) - 1.76}{6.02} \Rightarrow 9.7 \text{ bits}$$

But only the highest 6 bits contain valid signal measurement information.

$$ENOB_{signal} = \frac{S_{max,dB} - 1.76}{6.02} = \frac{40 - 1.76}{6.02} \Rightarrow 6.3 \text{ bits}$$

There are no fractional bits; the effective resolution is rounded down to the next integer.

---

[1] As opposed to the minimum *detectable* signal which should be much lower. The minimum measure signal is defined by the acceptable error tolerance.

If one LSB is defined as 192 nV, a 10-bit ADC full-scale range (FSR) would be 197 µV. Linear scaling to a more reasonable FSR — say, 1V — leads to the need for gain of about 5086 (~ 74 dB). This is a healthy amount of gain to provide in a single stage ... but IAs are available with gains into 1000s — at low bandwidths.

One potential and reasonable configuration would define an instrumentation amplifier gain of 508.6 (~54 dB) and an active filter gain of 10. This filter should be a Butterworth or Chebyshev II type if amplitude is a primary concern; a Thomson/Bessel type if phase or group delay is an issue. Note that the -3dB point is an amplitude reduction of 71%; accurate amplitude measures require a signal bandwidth to be a factor of 5 or more lower than the filter bandwidth.

But this booklet is not intended to be a discussion of application ... the instrumentation amplifier is a critical element in maintaining signal integrity; the question to be answered is how much error will the amplifier network itself add to the signal?

Chapter 1: The Instrumentation Amplifier

The circuit to be discussed:

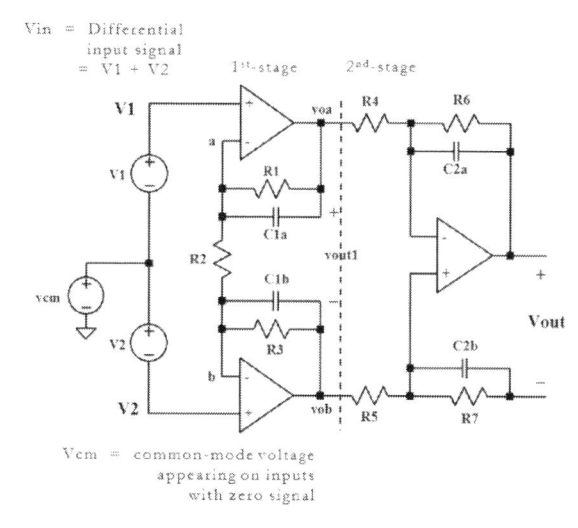

Figure 1: Discrete 3-Opamp Instrumentation Amp

Such an input signal might appear as:

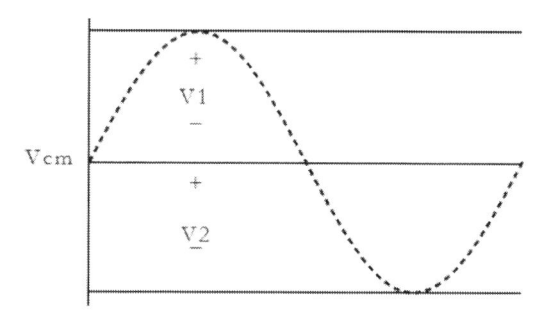

Figure 2: Illustrative relationship of vcm, V1, and V2

The common-mode voltage is often assumed to be zero volts, or "ground" (GND). An amplifier's compliance range is a measure of the range of vcm in which the amplifier will operate as expected.

For the analysis, one may generally assume identical amplifiers on a common substrate, no substrate leakage current, and all elements of the network at identical temperature.

However, this analysis is intended for a discrete level implementation; such assumptions may not be valid in practice ... but they form a solid basis for understanding the effects of errors.

The analysis will consist of two parts: the first will discuss the operation of the first – differential – stage; the second will discuss the differential to single-ended conversion stage.

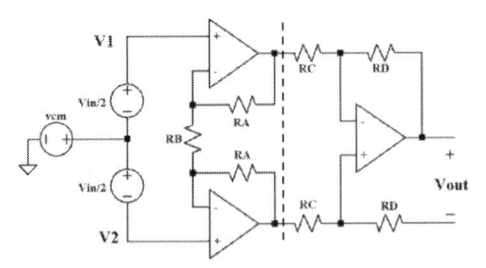

Figure 3: Discrete 3-Opamp Instrumentation Amp

It is known that the gain function of this particular topology is:

$$V_{outIA} = (V1 - V2)\left(1 + \frac{2\,RA}{RB}\right)\frac{RD}{RC}$$

when:
$R1 = R3 = RA$; $R2 = RB$
$R4 = R5 = RD$; $R6 = R7 = RC$

But sometimes these equalities don't hold true - and the amplifiers add errors of their own.

This analysis gets deeper into the details ...

Chapter 2: Symbolic Equations

Differential & Common-Mode Gain Error

Instrumentation amplifiers are often used for signals of near-DC frequency. Many earth-phenomena of interest are of very low frequency – how fast does temperature change? Next to the detection element itself, the low-frequency performance at this stage is often the limiting factor in effective accuracy of the measure. More bits may increase resolution but will not increase the inherent limits of signal accuracy.

The first effects to be discussed are those at zero-frequency; signals of finite frequency are affected by these and other effects such as unity-gain frequency, common-mode rejection ratio (CMRR), and power-supply rejection ratio (PSRR).

Network Stage 1

The instrumentation amplifier is best analyzed by breaking it into two stages: the dual amplifier difference stage and the single-amplifier gain stage. The difference stage is shown in Figure 4.

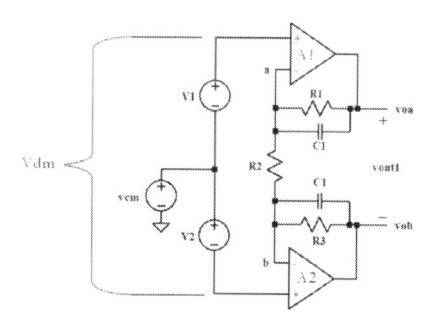

Figure 4: Stage 1 - Difference Stage of Instrumentation Amp

The capacitors provide frequency roll-off in the feedback path to aid in noise reduction. The capacitor values should be set to produce a pole at roughly 20x the maximum signal frequency.

Stage 1 Differential Gain

Assume vcm = 0; vin = vdm = V1 − V2; $A_{OL} \to \infty$

Under these assumptions, there will be no voltage swing between the input terminals. The summing junctions of A1 and A2 will be at potentials of V1 and V2 respectively.

The outputs of A1 and A2 will be the sum of the individual inputs.

For amplifier A1 via superposition:

$$V_{oa(V1)} = V_1 \frac{(R1 + R2)}{R2} \qquad V_{oa(V2)} = -V_2 \frac{R1}{R2}$$

$$V_{oa} = V_{oa(V1)} + V_{oa(V2)} \Rightarrow V_1 \frac{(R1 + R2)}{R2} + \left(-V_2 \frac{R1}{R2}\right)$$

For amplifier A2:

$$V_{ob(V1)} = -V_2 \frac{R3}{R2} \qquad V_{ob(V2)} = V_1 \frac{(R2 + R3)}{R2}$$

$$V_{ob} = V_{ob(V1)} + V_{ob(V2)} \Rightarrow \left(-V_2 \frac{R3}{R2}\right) + V_1 \frac{(R2 + R3)}{R2}$$

The output voltage is the difference between the two:

$$V_{out1} = V_{oa} - V_{ob} \Rightarrow (V1 - V2) \frac{R1 + R2 + R3}{R2}$$

with Vin = V1 − V2

Re-arranging terms, the 1st-stage differential gain is:

$$\boxed{A_{DM1} = \frac{V_{out1}}{Vin} = 1 + \frac{R1 + R3}{R2}}$$

Chapter 2: Symbolic Equations 11

The usual configuration sets R1 = R3 = Ra such that:

$$A_{DM1} = 1 + \frac{2\,Ra}{R2}$$

If R2 is also set equal to Ra as in this example:

$$A_{DM1} = 1 + \frac{2\,Ra}{Ra} = 3$$

R2 is a common-mode device; its effect is shared equally by both sides of the differential topology. As such, it may be used to set the gain with equal effect on both amplifiers. It also creates a handy tap point for active common-mode shielding: split the device into two equal value devices and use the common junction to drive a cable shield. One form of this connection is shown in the Appendix.

Although this is a discussion on DC analysis, it may be useful to consider the effect of the feedback capacitors.

The elements R1 and R3 in the describing expressions may be replaced with impedances:

$$A_{DM1} = \frac{V_{out1}}{V_{in}} = 1 + \frac{Z1 + Z3}{R2}$$

where Z1 = Z3 as: $Z = \dfrac{R}{1 + s\,C\,R}$

The differential gain expression is now:

$$A_{DM1} = \frac{R1}{R2\,(1 + s\,C1\,R1)} + \frac{R3}{R2\,(1 + s\,C3\,R3)}$$

With R1 = R3 = R2 = R and C1 = C3 = C

$$A_{DM1} = 1 + \frac{2}{1 + s\,C\,R}$$

Assume resistance Ra = 10k. The 1st-stage frequency response will be defined by the capacitors – assuming the closed-loop gain frequency corner of the amplifier is sufficiently greater than the corner frequency defined by the RC combination.

This is the assumption in the following frequency plots where Ra = 10k and C = {100 pF, 1000 pf, 0.01 µF, 0.1 µF, 0.1 µF}.

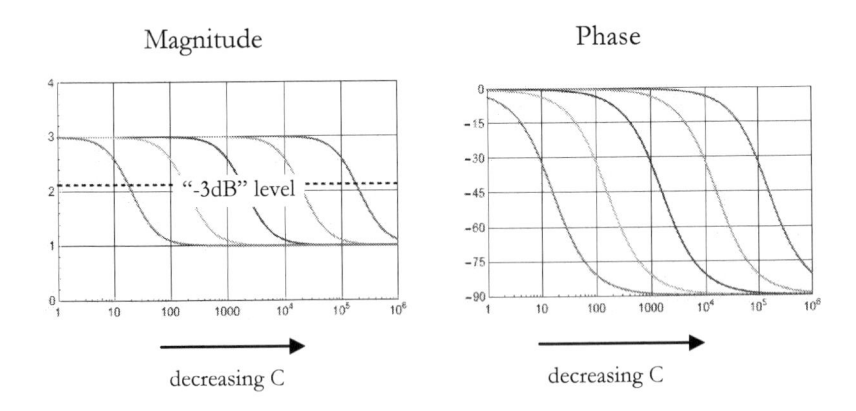

Magnitude

Phase

decreasing C

decreasing C

The signal bandwidth increases as the feedback capacitance decreases.

The effect of the capacitances helps limit noise contributions of the 1st-stage – which typically provides the highest gain of the system.

Even though not discussed further herein, frequency and phase response can become critical even at near-zero frequencies.

Chapter 2: Symbolic Equations 13

Stage 1 Common-Mode Gain

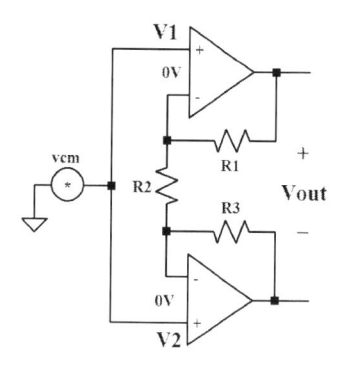

Assume Vin = 0.
Therefore, V1 = V2 = vcm

If A1 and A2 have very high, matched open-loop gains, the potential on both terminals of R2 will be Vcm, therefore, IR2 = 0. Amplifier mismatch will create an offset effect which needs to be considered when determining common-mode rejection.

If the opamp bias currents are assumed negligible, the resistor currents are zero: $I_{R1} = I_{R2} = I_{R3} = 0$.

Therefore voa = vob = V1 = V2 = vcm

These ideal conditions lead to an ideal expression:

$$A_{CM1} = \frac{V_{out1}}{vcm} = \frac{voa - vob}{vcm} = \frac{vcm - vcm}{vcm} = 0$$

Life would be easier if it were so ...

The two amplifiers will never be perfectly matched. Finite open-loop gain and resistor mismatch are the primary degradation factors. Temperature variations may need to be considered in a physical network.

For finite-gain amplifiers, each having open-loop gains of A_{OL}:

$$voa_{(cm)} = Vcm \frac{1 + \frac{R1}{R2}}{1 + \frac{1}{A_{OL1}}\left(1 + \frac{R1}{R2}\right)}$$

$$voa_{(V2)} = -V2 \frac{R1}{R2} \frac{1}{1 + \frac{1}{A_{OL1}}\left(1 + \frac{R1}{R2}\right)}$$

A little manipulation gives results for the output of amplifier A1:

$$\text{voa} = \text{voa}_{(cm)} + \text{voa}_{(V2)} = \frac{\text{Vcm}(R1 + R2) - V2\,R1}{R2\left[1 + \frac{1}{A_{OL1}}\left(1 + \frac{R1}{R2}\right)\right]}$$

Similarly for the output of amplifier A2:

$$\text{vob}_{(cm)} = \text{vcm}\,\frac{1 + \frac{R3}{R2}}{1 + \frac{1}{A_{OL2}}\left(1 + \frac{R3}{R2}\right)}$$

$$\text{vob}_{(V2)} = -V2\,\frac{R3}{R2}\,\frac{1}{1 + \frac{1}{A_{OL2}}\left(1 + \frac{R3}{R2}\right)}$$

... which provides a solution for vob ...

$$\text{vob} = \text{vob}_{(cm)} + \text{vob}_{(V2)} = \frac{\text{Vcm}(R2 + R3) - V1\,R3}{R2\left[1 + \frac{1}{A_{OL2}}\left(1 + \frac{R3}{R2}\right)\right]}$$

These expressions may be re-arranged:

$$\text{voa} = A_{OL1}(\text{Vcm} - V1) \Rightarrow V1 = \text{Vcm} - \frac{\text{voa}}{A_{OL1}}$$

$$\text{vob} = A_{OL2}(\text{Vcm} - V2) \Rightarrow V2 = \text{Vcm} - \frac{\text{vob}}{A_{OL2}}$$

Chapter 2: Symbolic Equations   15

The differential output of the first stage is:

$$voa = \frac{Vcm(R1 + R2) - R1\left(Vcm - \dfrac{vob}{A_{OL2}}\right)}{R2\left[1 + \dfrac{1}{A_{OL1}}\left(1 + \dfrac{R1}{R2}\right)\right]}$$

$$vob = \frac{Vcm(R3 + R2) - R3\left(Vcm - \dfrac{voa}{A_{OL1}}\right)}{R2\left[1 + \dfrac{1}{A_{OL2}}\left(1 + \dfrac{R3}{R2}\right)\right]}$$

Equating terms::

$$voa\, R2\left[1 + \frac{1}{A_{OL1}}\left(1 + \frac{R1}{R2}\right)\right] - vob\,\frac{R1}{A_{OL2}} \equiv Vcm\, R2$$

$$vob\, R2\left[1 + \frac{1}{A_{OL2}}\left(1 + \frac{R3}{R2}\right)\right] - voa\,\frac{R3}{A_{OL1}} \equiv Vcm\, R2$$

Two equations; two common unknowns.

One method of solution is to define factors k1 and k2:

$$k_1 \equiv \left[1 + \frac{1}{A_{OL1}}\left(1 + \frac{R1}{R2}\right)\right] \qquad k_2 \equiv \left[1 + \frac{1}{A_{OL2}}\left(1 + \frac{R3}{R2}\right)\right]$$

Substitute terms:

$$voa\, R2\, k1 - vob\,\frac{R1}{A_{OL2}} \equiv Vcm\, R2$$

$$-voa\,\frac{R3}{A_{OL1}} + vob\, R2\, k2 \equiv Vcm\, R2$$

Apply a little algebraic magic:

$$voa = vcm \frac{R2\,k2\,A_{OL2} + R1}{k2\,A_{OL2}} \cdot \frac{R2\,k2\,A_{OL1}\,A_{OL2}}{R2^2\,k1\,k2\,A_{OL1}\,A_{OL2} - R1\,R3}$$

$$\Rightarrow vcm \frac{R2\,A_{OL1}(R2\,k2\,A_{OL2} + R1)}{R2^2\,k1\,k2\,A_{OL1}\,A_{OL2} - R1\,R3}$$

and:
$$vob = vcm \frac{R2\,A_{OL2}(R2\,k1\,A_{OL1} + R3)}{R2^2\,k1\,k2\,A_{OL1}\,A_{OL2} - R1\,R3}$$

The 1st-stage common-mode gain is defined as: $A_{CM1} = \dfrac{voa - vob}{vcm}$

Substituting from above:

$$A_{CM} = \frac{\left(Vcm\,\dfrac{R2\,A_{OL1}(R2\,k2\,A_{OL2}+R1)}{R2^2\,k1\,k2\,A_{OL1}\,A_{OL2}-R1R3}\right) - \left(Vcm\,\dfrac{R2\,A_{OL2}(R2\,k1\,A_{OL1}+R3)}{R2^2\,k1\,k2\,A_{OL1}\,A_{OL2}-R1R3}\right)}{Vcm}$$

$$= \frac{R2\,A_{OL1}(R2\,k2\,A_{OL2}+R1) - R2\,A_{OL2}(R2\,k1\,A_{OL1}+R3)}{R2^2\,k1\,k2\,A_{OL1}\,A_{OL2} - R1\,R3}$$

gives one form of expression for the stage 1 common-mode gain:

$$A_{CM1} = \frac{A_{OL1}(A_{OL2} + \Re) - A_{OL2}(A_{OL1} + \Re)}{\left(A_{OL1} + \dfrac{R1}{R2}\right) + \left(A_{OL2} + \dfrac{R3}{R2}\right) - \dfrac{R1\,R3}{R2^2}}$$

where $\Re = \dfrac{R1 + R3}{R2}$

Reducing the expression gives an equivalent expression:

$$\boxed{A_{CM1} = \frac{(R1+R3)(A_{OL1} - A_{OL2})}{R1\,A_{OL2} + R2\,A_{OL1}\,A_{OL2} + R3\,A_{OL1}}}$$

Network Stage 2

2nd-Stage Difference Amplifier Gain

The 2nd-stage of the instrumentation amplifier is shown schematically as:

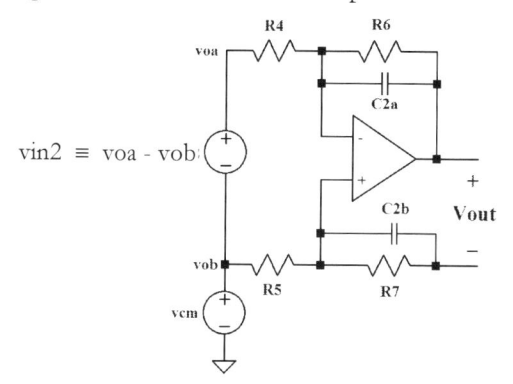

Figure 5: Stage 2 – Differential-to-Single End Converter

In this configuration, vob = vcm , voa = vin2 + vcm.

The bias currents and offset voltage are negligible.

The common-mode voltage of the 2nd-stage vcm2 will include any offset voltage of the 1st-stage.

This configuration is a basic single-stage difference amplifier which may be represented with a more complex impedance model of form:

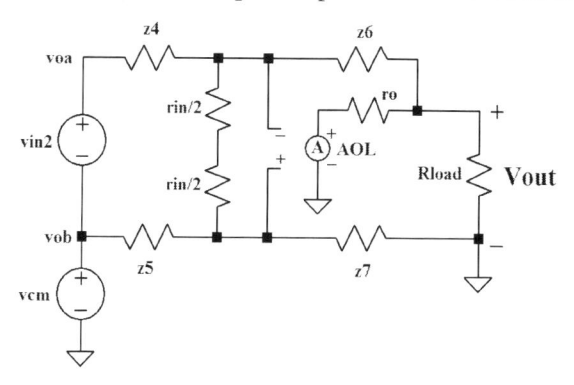

Figure 6: Stage 2 – Thevenin Equivalent

Define a few notation terms for presentation simplicity.

$$ziz4 = zi \| z4 = \frac{zi\, z4}{zi + z4} \qquad ziz6 = zi \| z6 = \frac{zi\, z6}{zi + z6}$$

$$zoz4 = zo \| z1 = \frac{zo\, z4}{zo + z4} \qquad zoz6 = zo \| z6 = \frac{zo\, z6}{zo + z6}$$

$$z5z7 = z5 \| z7 = \frac{z5\, z7}{z5 + z7}$$

and $zi \rightarrow zin + z5z7$

Since vcm is set to zero in superposition analysis for voa, z5 is parallel to z7 and the combination in series with amplifier input impedance zi. The Thevenin network is simplified to:

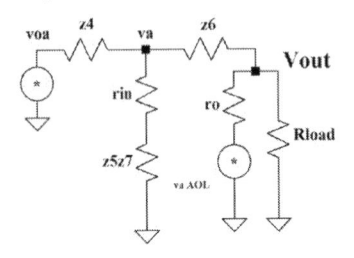

Figure 7: Stage 2 – Reduced Thevenin Equivalent

with the following network expressions:

$$va = voa \frac{ziz6}{z4 + ziz6} + vout \frac{ziz4}{z6 + ziz4}$$

$$vout = va \frac{-aol\, z6 + zo}{z4 + ziz6} + vout \frac{zoz6}{zload}$$

Re-arranging expressions to eliminate term va:

$$vout = voa \frac{zi\, z6\, zload\, (zo - aol\, z6)}{[z6\, zi + z4(z6 + zi)]\, zload\, (z4 + zo)}$$

$$+\; vout \frac{zo\, [z6^2\, zi + z4\, (z6^2 + z6\, zi + zi\, zload)] - aol\, z4\, z6\, zi\, zload}{[z6\, zi + z4(z6 + zi)]\, zload\, (z4 + zo)}$$

Chapter 2: Symbolic Equations

Re-arranging terms to obtain the network transfer function:

$$\frac{vout}{voa} = \frac{zi\, zload\,(zo - aol\, z6)}{(z4 + zi)[z6\,(zload - zo) + zload\, zo] + z4\, zi\,(zload\,(1 + aol) - zo)}$$

If $zi \rightarrow \infty$

$$\frac{vout}{voa} = \frac{zi\, zload\,(zo - aol\, z6)}{z6\,(zload - zo) + z4\,(zload\,(1 + aol) - zo) + zload\, zo}$$

If $zo \rightarrow 0$

$$\frac{vout}{voa} = -\frac{aol\, z6\, zi\, zload}{z6\, zi\, zload + z4\,[z6\, zload + z4\, zi\,(zload\,(1 + aol))]}$$

If $zo \rightarrow 0$ and $zi \rightarrow \infty$

$$\frac{vout}{voa} = -\frac{aol\, z6\, zload}{z6\, zload + z4\,[zload\,(1 + aol)]}$$

If $zload \rightarrow \infty$

$$\frac{vout}{voa} = -\frac{aol\, z6}{z6 + z4\,(1 + aol)}$$

and if $aol \rightarrow \infty$

$$\frac{vout}{voa} = -\frac{z6}{z4}$$

Consider one of the parallel combinations. The usual assumption is that rin tends to infinity – it certainly makes the math easier ... but consider:

Let rin = 100M and $z4$ = 10k. The parallel combination is:

$$zi \parallel z4 = \frac{zi\, z4}{zi + z4} \Rightarrow \frac{100 \times 10^6 \times 10 \times 10^3}{100 \times 10^6 + 10 \times 10^3} = 9999.0\, \Omega$$

Many if not most modern opamps have input impedances greater than 100 M but it is always a good idea to confirm this.

This analysis assumes ro of zero and ri of ∞.

2nd-Stage Differential and Common-Mode Gain

The non-inverting input node voltage vp defines the inverting input node voltage vm where vm = vp.

$$vp = vcm \frac{R7}{R5 + R7}$$

Since the currents through R4 and R6 are equal:

$$I4 = \frac{vin - vm}{R4} \Leftrightarrow I6 = \frac{vm - vout}{R6}$$

Substituting for vm:

$$\frac{vin - vm}{R4} = \frac{vm - vout}{R6} \Rightarrow \frac{vin}{R4} + \frac{vout}{R6} = vm \left(\frac{1}{R6} + \frac{1}{R4} \right)$$

or:

$$\frac{vin}{R4} + \frac{vout}{R6} = \left(vcm \frac{R7}{R5 + R7} \right) \left(\frac{1}{R6} + \frac{1}{R4} \right)$$

Solving this expression for vout gives

$$V_o = vin \left(-\frac{R6}{R4} \right) + vcm \left(\frac{R4\, R7 - R5\, R6}{R4(R5 + R7)} \right)$$

The differential-mode gain is:

$$\boxed{V_{odm2} \Big|_{vcm = 0} \equiv vin \left(-\frac{R6}{R4} \right)}$$

The common-mode gain is:

$$\boxed{V_{ocm2} \Big|_{vin = 0} \equiv vcm \left(\frac{R4\, R7 - R5\, R6}{R4(R5 + R7)} \right)}$$

With ideal matching: $A_{CM2} \to 0$ for $\frac{R6}{R4} = \frac{R7}{R5} \Rightarrow 0$

The 2nd-stage resistance mismatch defines the common-mode gain.

The matching of the 2nd-stage resistors determines the quality of overall common-mode rejection.

Chapter 2: Symbolic Equations

Effect of Resistor Mismatch on ACM2

For a set of discrete resistors having identical characteristics in a common operation environment, a tolerance parameter may be defined which applies to all.

Define the resistors in terms of tolerance k. R5 and R7 are defined in terms of R4 and R6:

$$R4_{min} = R4\,(1-k) \qquad R6_{max} = R6\,(1+k)$$

$$R5_{max} = R5\,(1+k) \qquad R7_{min} = R7\,(1-k)$$

where parameter k represents the worst-case tolerance of the resistors.

Define a scaling parameter n. This parameter applies to feedback resistors R5 and R7 as scaled values of R4 and R6.

$$R5_{max} = n\,R4\,(1+k) \qquad R7_{min} = n\,R6\,(1-k)$$

Terms R5 and R7 fall out of the expression.

Parameter "n" is assumed equal to 1 where R5 = R4 and R7 = R6.

2nd-stage common-mode gain with equal resistor tolerance:

$$\boxed{A_{CM2wc} = \frac{-4k\,R6}{(1-k)\,[\,R4\,(1+k) + R6\,(1-k)\,]}}$$

This expression is valid for resistors having identical tolerances.

If the individual tolerances vary, the 2nd-stage common-mode gain with distributed resistance tolerance:

$$\boxed{A_{CM2wc} = \frac{R4\,(1-k4)\,R7\,(1-k7) - R5\,(1+k5)\,R6\,(1+k6)}{R4\,(1-k4)\,[\,R5\,(1+k5) + R7\,(1-k7)\,]}}$$

Offset Voltage and Bias Current Error

Offset Voltage Error: ε_{os}

Using this network as a model:

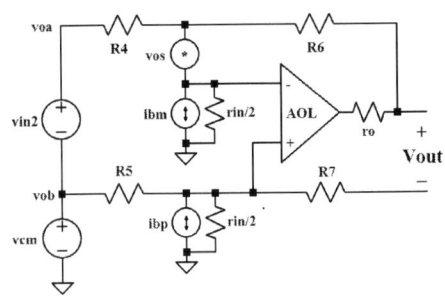

Figure 8: 2nd-Stage DC Error Sources

Resistance rin is the differential input resistance of the amplifier, vos is the input offset voltage, ibm and ibp are the input bias currents. The output offset error is found by setting vcm to 0, placing vob to GND.

Offset voltage is defined as the voltage required at node a to force Vout to 0 with both nodes b and c at GND potential. Bias current is defined as the current required at node a to force Vout to 0 with node b at GND potential and node c open. Thevenin equivalency is used to derive the appropriate expressions.

Since Vin = Vout = 0 and ibm = 0, the following networks may be used for Thevenin reduction:

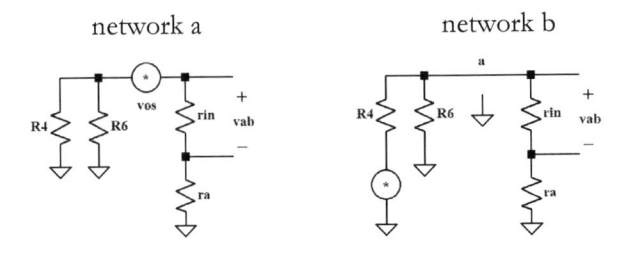

Figure 9: Offset Voltage Error Reduction

Chapter 2: Symbolic Equations

The Thevenin network expressions for the two networks are:

network a

$$V_{ab} = vos \frac{rin}{rin + ra + req}$$

$$ra = \frac{R5\ R7}{R5 + R7}$$

$$req = \frac{R4\ R6}{R4 + R6}$$

network b

$$V_{eq} = \varepsilon os \frac{R6}{R4 + R6}$$

$$req = \frac{R4\ R6}{R4 + R6}$$

which can be reduced to an equivalent network of:

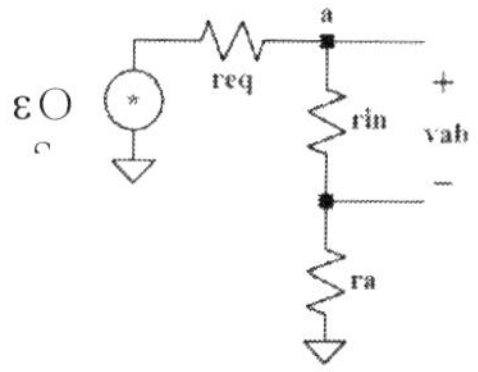

$$vab = veq \frac{rin}{rin + ra + req}$$

$$= \varepsilon os \frac{R6}{R4 + R6} \frac{rin}{rin + ra + req}$$

Figure 10: Offset Voltage Error Reduction

Equating expressions for vab gives the expression for total offset voltage error εos:

$$ve \frac{R6}{R4 + R6} \frac{rin}{rin + ra + req} \equiv vos \frac{rin}{rin + ra + req}$$

which results in the expression:

$$\boxed{\varepsilon os = vos \frac{R4 + R6}{R6}}$$

Note the offset error voltage εos is independent of rin, R5, and R6.

Inverting Bias Current Error: εibm

The bias current is reflected back to the input as an equivalent voltage εibm:

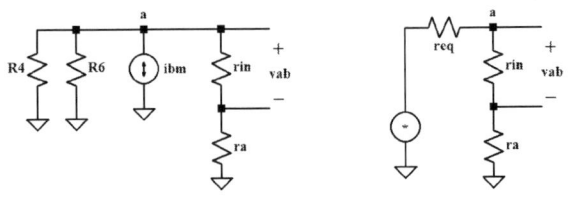

Figure 11: Inverting Bias Current Error Reduction

The Thevenin network expressions:

$$a = ibm \frac{req\ (ra + rin)}{rin + ra + req} \qquad a = veq \frac{(ra + rin)}{rin + ra + req}$$

$$req = \frac{R4\ R6}{R4 + R6} \qquad ra = \frac{R5\ R7}{R5 + R7} \qquad veq = \varepsilon ibm \frac{R6}{R4 + R6}$$

Equating these expressions:

$$\varepsilon ibm \frac{R7}{R4 + R7} \frac{rin + ra}{rin + ra + req} \Rightarrow \varepsilon ibm \frac{R7(rin + ra)}{rin + ra + req}$$

$$\Rightarrow ibm \frac{req\ (ra + rin)}{rin + ra + req} \equiv \varepsilon ibm \frac{R7(rin + ra)}{rin + ra + req}$$

Shuffling terms about a bit more ...

$$\varepsilon ibm = ibm \frac{req\ (R4 + R7)}{R7} = ibm \frac{R4\ R7}{R4 + R7} \frac{(R4 + R7)}{R7}$$

The 2nd-stage error due to the inverting input bias current:

$$\boxed{\varepsilon ibm = ibm\ R4}$$

The bias equivalent error voltage εibm is not dependent on R5, R6, R7, or rin (where rin is the often very high input resistance of the opamp).

Chapter 2: Symbolic Equations

Bias Current Offset Error: εibp

Parameter ibp is the bias current into the +input of the amplifier. A similar process as above will permit the representation of positive bias current as a voltage ibp. This tends to offset the bias current error.

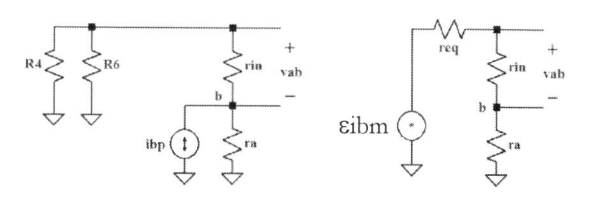

Figure 12: Non-Inverting Bias Current Error Reduction

The expression for bias current error in the non-inverting input:

$$vb = ibp\ ra \qquad ra = \frac{R5\ R7}{R5 + R7}$$

vb is reflected to the output as:

$$vb' = ibp\ ra\ \frac{R4 + R6}{R7}$$

This is reflected back to the input as:

$$\varepsilon ibm = ibm\ \frac{req\ (R4 + R6)}{R6} = ibm\ \frac{R4\ R6}{R4 + R6}\ \frac{(R4 + R6)}{R6}$$

$$\varepsilon ibp = -vb'\ \frac{R4}{R6} \Rightarrow -ibp\ ra\ \frac{R4 + R6}{R6}\ \frac{R4}{R6}$$

Giving the expression for the 2nd-stage error due to the non-inverting input bias current:

$$\boxed{\varepsilon ibp = -ibp\ ra\ \frac{(R4 + R6)}{R6} \Rightarrow -ibp\ \frac{R5\ R7}{R6}\ \frac{R4 + R6}{R5 + R7}}$$

Total 2nd-Stage DC Error Model

All the 2nd-stage DC offset/bias components may be defined with the resulting model:

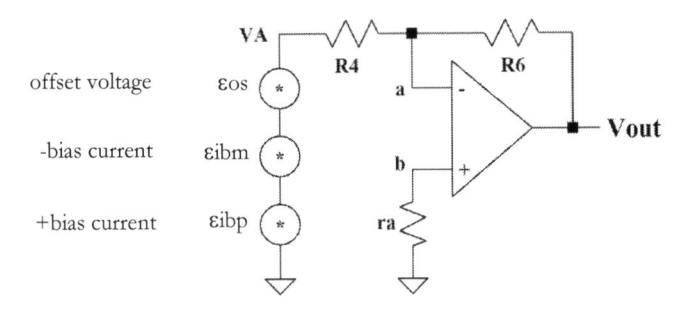

Figure 13: 2nd-Stage DC Error

where the cumulative error voltages are:

$$\varepsilon os = vos \frac{(R4 + R6)}{R6}$$

$$\varepsilon ibm = ibm \, R4 \qquad \varepsilon ibp = -ibp \, \frac{R5 \, R7}{R6} \, \frac{R4 + R6}{R5 + R7}$$

The current ibp flowing from the +input will produce a voltage drop across ra which tends to cancel the error caused by the current is. This causes .ibp to have a negative sign, thereby reducing the effect of εibm.

It is often the case that bias currents ibm = ibp; the bias current offset error is then related to resistor mismatch.

For identical resistors:

$$\varepsilon ibm + \varepsilon ibp = ibm \, R4 + \left(-ibp \, \frac{R5 \, R7}{R6} \, \frac{R4 + R6}{R5 + R7} \right)$$

$$= ibm \, R - ibp \, R$$

Chapter 2: Symbolic Equations

Given resistance values with tolerances as before:

$$R4_{min} = R4_o(1-k) \qquad R6_{max} = R6_o(1+k)$$

$$R5_{max} = n\, R5_o(1+k) \qquad R7_{min} = n\, R7_o(1-k)$$

with equal bias currents ibp = ibm:

$$\Delta\varepsilon ib = \varepsilon ibm + \varepsilon ibp$$

$$= ib(1-k)\left(R4 - n\,\frac{R5\,R7}{R6}\,\frac{R4(1-k)+R6(1+k)}{R5(1+k)+R7(1-k)}\right)$$

... and if the resistances are also identically equal:

$$\varepsilon ibm - \varepsilon ibp = ib\,\Re\,(k-1)(n-1) \;\Rightarrow\; 0 \text{ when } n=1$$

The total DC error of the 2nd-stage is therefore expressed as:

$$V_{o,DCerr} = -\frac{R6}{R4}\left[\varepsilon os + |\varepsilon ibm - \varepsilon ibp|\right]$$

or the full expression for DC error:

$$\boxed{V_{o,DCerr} = -\frac{R6}{R4}\left[vos\,\frac{R4+R6}{R6} + \left|ibm\,R4 - ibp\,\frac{R5\,R7}{R6}\,\frac{R4+R6}{R5+R7}\right|\right]}$$

Chapter 3: Total System Error

System Differential Gain

This is the factor referred to as the "gain" of the amplifier where:

$$A_{DM} = A_{DM1} A_{DM2}$$

The gain of the system is defined by the multiplication of the differential gains of each stage:

$$A_{DM} = \left| \left[1 + \frac{R1 + R3}{R2} \right] \times \left[-\frac{R6}{R4} \right] \right|$$

If all resistors are equal, the inherent amplifier gain is $A_{DM} = 3$

Note the "polarity" depends on which input is defined "+". Unlike an opamp, both instrumentation amplifier inputs are high-impedance, independent, and truly differential. The amplifier does not require an "external" feedback network.

The inputs may be positive or negative with respect to each other; the terminal designation defines the mathematical operation: the magnitude of the minus input is subtracted from the magnitude of the positive input.

$$V_{out} = (V_1 - V_2) A_{DM}$$

Chapter 3: Total System Error

System Common-Mode Gain:

The common-mode voltage at the output of the 1st-stage is (Va − Vb). The 2nd-stage sees this voltage as part of the differential signal. The overall common-mode voltage at the 2nd-stage output will be:

$$V_{ocm} = (Va + Vb)\frac{R6}{R4} + V_{cm} A_{CM2}$$

The definition of 1st-stage common-mode gain was given as:

$$A_{CM1} = \frac{Va + Vb}{vcm} \rightarrow Va + Vb = V_{cm} A_{CM1}$$

which gives a result for the amplifier common-mode gain of:

$$A_{CM} = A_{CM1}\frac{R6}{R4} + A_{CM2}$$

where, from the descriptive expressions:

$$A_{CM1} = \frac{R1\,R3\,(A_{OL1} - A_{OL2})}{R1\,A_{OL1} + R2\,A_{OL1}\,A_{OL2} + R3\,A_{OL1}}$$

$$A_{CM2} = \frac{R4\,R7 - R5\,R6}{R4\,(R5 + R7)}$$

However, the actual determination of common-mode gain will be dependent on both resistor tolerances and the effects of the amplifiers.

Total DC Offset Error

The total DC offset and bias error is expressed as:

$$V_{0,DCerr} = -\frac{R6}{R4}\left[vos\frac{(R4+R6)}{R6} + \left|ibm\ R4 - ibp\frac{R5\ R7}{R6}\frac{R4+R6}{R5+R7}\right|\right]$$

This error may be nulled during a calibration process. A problem arises when environmental drift effects are accounted for.

These expressions form the basis for numerical results.

The resistors are often the most sensitive elements of the network; they need to be considered in some detail.

Common-Mode Rejection Ratio (CMRR)

The CMRR is determined by the ratio of differential-mode gain divided by the common-mode gain:

$$CMRR = \frac{A_{DM}}{A_{CM}}$$

Substituting terms, the CMRR is found to be expressed as:

$$CMRR = \frac{\left|\left[1+\frac{R1+R3}{R2}\right]\left[-\frac{R6}{R4}\right]\right|}{A_{CM1}\frac{R6}{R4} + A_{CM2}}$$

of whose presentation gets sort of ugly to place on a narrow page when the values for common-mode gain are expanded ...

$$A_{CM1} = \frac{R1\ R3\left(A_{OL1} - A_{OL2}\right)}{R1\ A_{OL1} + R2\ A_{OL1}\ A_{OL2} + R3\ A_{OL1}}$$

$$A_{CM2} = \frac{R4\ R7 - R5\ R6}{R4\ (R5 + R7)}$$

The resistors characteristics are necessary to determine a numerical value of the network CMRR.

Chapter 3: Total System Error

Resistor Tolerance & Aging
Tolerance
1st-Stage Network

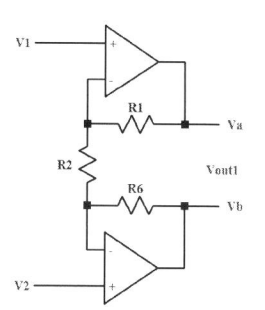

The gain expressions have been previously determined to be:

$$A_{DM1} = 1 + \frac{R1 + R3}{R2}$$

and:

$$A_{CM1} = \frac{R1\,R3\,(A_{OL1} - A_{OL2})}{R1\,A_{OL2} + R2\,A_{OL1}\,A_{OL2} + R3\,A_{OL1}}$$

Figure 14: 1st-Stage Resistor Mismatch Error

For discrete resistors, the worst-case extremes are:

$$A_{DM1,max} = 1 + \frac{R1(1 \pm k) + R3(1 \pm k)}{R2(1 \mp k)}$$

$$A_{CM1} = \frac{R1\,R3\,(1 \pm k)^2\,(A_{OL1} - A_{OL2})}{(1 \pm k)[R1\,A_{OL2} + R3\,A_{OL1}] + R2\,(1 \mp k)\,A_{OL1}\,A_{OL2}}$$

where k is the resistance tolerance (to be discussed later).

The tolerances of a matched array will track, but for the example numerical analysis, the arrays are 4-packs. Assuming one package is used for the 1st-stage, another for the 2nd-stage, it cannot be assumed that the tolerances of the two packs will track.

Note that R2 can be configured as a user-set device to define the gain of the 1st-stage. This is not an uncommon practice in integrated instrumentation amplifiers.

2nd-Stage Network

The worst-case resistor mismatch may be found for resistors of identical type in the following difference amplifier network:

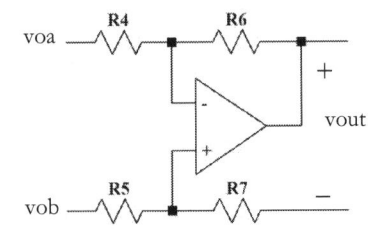

Figure 15: 2nd-Stage Resistor Mismatch Error

For a given total resistance tolerance k:

$$R4_{min} = R4\,(1-k) \qquad R5_{max} = n\,R4\,(1+k)$$
$$R6_{max} = R6\,(1+k) \qquad R7_{min} = n\,R6\,(1-k)$$

where "n" is a resistance scaling factor set to 1 for equal resistance (difference gain of 1).

The output voltage can be expressed as:

$$V_{out} = \frac{R6}{R4}\left(vob\,\frac{R4(1-k) + R6(1+k)}{R4(1+k) + R6(1-k)} - voa\,\frac{(1+k)}{(1-k)}\right)$$

where, if tolerance $k = 0$:

$$V_{out} = -\frac{R6}{R4}(vob - voa)$$

The worst-case common-mode gain due to tolerance mismatch is therefore expressed:

$$A_{CM,min} = \frac{R4_{min}\,R7_{min} - R5_{max}\,R6_{max}}{R4_{min}\,(R5_{max} + R7_{min})}$$

Aging

The Arrhenius equation describes the effects of temperature and age on reaction rates. This is an empirical expression but has been shown to be a valid approximation for resistors. The rule-of-thumb" suggests a doubling of nominal rate for every 10°C increase in temperature. This may be expressed as:

$$A_T = A_{To} \; 2 \wedge \left[\frac{T - T_o}{10} \right]$$

Consider the specifications for a typical precision thin-film resistor:

"Load life" is ±0.1% after 10k hrs. "Shelf" life is 50 ppm/yr at room temperature. Maximum power rating is given as 300 mW at 70°C.

What is the decrease in tolerance at a temperature of 60°C?

At T = 70°C:

$$A_{70} = 0.1\% \text{ per } 10k \text{ hr} \Rightarrow 1000 \text{ ppm}/10000 \text{ hr}$$
$$= 0.1 \text{ ppm/hr}$$

At To = 25°C:

$$A_{25} = 50 \text{ ppm/yr} \Rightarrow 50 \text{ ppm}/8760 \text{ hr}$$
$$= 0.0057 \text{ ppm/hr}$$

Since the common rule-of-thumb for aging effects suggests an increase by roughly a factor $\delta = 2$ for every 10°C increase:

$$A_{60} = 0.0057 \times 2 \wedge \left[\frac{70 - 25}{10} \right] = 64.5 \text{ ppm/hr} \Rightarrow 565 \text{ ppm/yr}$$

However, "rules-of-thumb" should be confirmed for anything more than "rule-of-thumb" calculations..

The aging coefficient δ may be determined from:

$$\frac{A_T}{A_{T_o}} = \delta \wedge \left[\frac{T-T_o}{10}\right] \Rightarrow \delta = \frac{A_T}{A_{T_o}} \left[\frac{10}{T-T_o}\right]$$

Using the values given:

$$\frac{A_{70}}{A_{25}} = \delta \wedge \left[\frac{70-25}{10}\right] \Rightarrow \delta = \frac{A_{70}}{A_{25}} \left[\frac{10}{70-25}\right]$$

$$\Rightarrow \delta = \frac{0.1}{0.0057}^{[0.222]} = 1.889$$

The maximum component ambient temperature is given as 60TC; the maximum aging rate is calculated as:

$$A_{60} = A_{25} \times 1.889 \wedge \left[\frac{60-25}{10}\right] \approx 463 \, \text{ppm}/\text{yr}$$

The approximation of $\delta = 2$ has more than 5% error in this example; it can be shown that this "rule-of-thumb" can lead to even greater errors.

Chapter 4: Numerical Example

Expression Summary
The basic symbolic network expressions have been determined to be:

1st-Stage:
EQ 1: 1st-Stage Differential Gain

$$A_{DM1} = \frac{V_{out1}}{V_{in}} = 1 + \frac{R1 + R3}{R2}$$

EQ 2: 1st-Stage Common-Mode Gain

$$A_{CM1} = \frac{(R1 + R3)(A_{OL1} - A_{OL2})}{R1\, A_{OL2} + R2\, A_{OL1}\, A_{OL2} + R3\, A_{OL1}}$$

2nd-Stage:
EQ 3: 2nd-Stage Differential Gain (ideal):

$$A_{DM2} = \frac{V_{out}}{V_{in}} \Rightarrow \left. -\frac{R6}{R4} \right|_{R7\,R5 = R4\,R6}$$

EQ 4: 2nd-Stage Common-Mode Gain

$$A_{CM2} \equiv \frac{V_{ocm}}{V_{CM}} \Rightarrow \frac{R4\,R7 - R5\,R6}{R4\,(R5 + R7)}$$

EQ4a: 2nd-Stage Common-Mode Gain – identical tolerance

$$A_{CM2} = \frac{-4k\,R6}{(1-k)\,[R4(1+k) + R6(1-k)]}$$

EQ4b: 2nd-Stage Common-Mode Gain – distributed tolerance

$$A_{CM2} = \frac{R4(1-k4)\,n\,R6(1-k6) - n\,R4(1+k5)\,R6(1+k6)}{R4(1-k4)\,[n\,R4(1+k5) + n\,R6(1-k7)]}$$

2nd-Stage Offset Voltage & Bias Current:

EQ 5: Offset Voltage Error

$$\varepsilon os = vos \frac{R4 + R6}{R6}$$

EQ6a: Inverting Input Bias Current Error:

$$\varepsilon ibm = ibm\, R4$$

EQ6b: Non-Inverting Input Bias Current Error:

$$\varepsilon ibp = -ibp\, \frac{R5\, R7}{R6}\, \frac{R4 + R6}{R5 + R7}$$

The two bias currents offset each other. The total effect of bias current error is expressed:

$$\varepsilon ibtot = \varepsilon ibm + \varepsilon ibp$$

1) when resistors are identical:

$$\varepsilon ibtot = ibm\, R4 + \left(-ibp\, \frac{R5\, R7}{R6}\, \frac{R4 + R6}{R5 + R7}\right)$$

$$= ibm\, R - ibp\, R$$

2) with identical bias current and resistor tolerance k:

$$\varepsilon ibtot = ib(1-k)\left(R4 - n\frac{R5\, R7}{R6}\, \frac{R4(1-k) + R6(1+k)}{R5(1+k) + R7(1-k)}\right)$$

3) with identical bias current and identical resistors:

$$\varepsilon ibm - \varepsilon ibp = ib\, \Re\, (k-1)(n-1) \Rightarrow 0 \text{ when } n = 1$$

EQ7: Total 2nd-Stage DC Error

$$V_{o,DCerr} = -\frac{R6}{R4}\left[\varepsilon os + |\varepsilon ibm - \varepsilon ibp|\right]$$

Chapter 4: Numerical Example 37

System Analysis

Operating Parameters

Define a system for operation in space.

28VDC ±6V supply voltage ; 7-year end-of-life (EOL).
Ambient temperature range of -15°C → 30°C.

Temperature Assumptions

Since this is a vacuum environment, all thermal dissipation will be conductive. Assume a 30°C gradient between baseplate and components.

The worst-case component temperature range will be:

$$\Delta T = -15°C \rightarrow 60°C = 75°C \text{ or } 25°C \ \{+35, -40\}.$$

Resistors: Vishay Precision Group Z-style thin-film.

Two types to be compared; identical material from same manufacturer:

Discrete RNC90Z

Matched Array PNRD

All resistors have nominal value of 10kΩ

Opamp: Intersil ISL70417SEH (quad)
This is assumed 4 opamps on a common-substrate. Matching characteristics and cross-channel leakage are not quantified.

$V_{OS,typ}$	=	10 µV	$V_{OS,max}$	=	75 µV	$V_{OS,max,ra}$	=110 µV
$I_{b,typ}$	=	80 pA	$I_{b,max}$	=	5 nA		
$I_{bos,typ}$	=	80 pA	$I_{bos,max}$	=	3 nA		
$I_{b,tc}$	=	5 pA/°C					
$CMRR_{typ}$	=	145 dB	$CMRR_{min}$	=	120 dB		
$A_{OL,min}$	=	300E6	$A_{OL,typ}$	=	1.4E9		
GBW	=	1.5 MHz					

The minimum common-mode rejection due to the amplifier:

$$A_{CM,amp} = log^{-1}\left(-\frac{120}{20}\right) = 1.0 \times 10^{-6}$$

Resistor Analysis

One might think that resistor analysis would be straight-forward ...

With today's manufacturing techniques, a variety of resistor types are available. For this example project, a high-precision device with high reliability for long-life in a radiation environment is required. These types are available as either discrete devices or as integrated matched arrays.

As an example, the performance over life of a matched resistor array and a discrete resistor of the same technology and manufacturer are compared. As this defines the ability of the amplifier to extract the desired differential-mode signal from the common-mode variations, care is taken in the detailed analysis.

The basic question: Does the performance of this structure outperform a manufactured instrumentation amplifier?

A primary differentiation criterion is often based on CMRR.

The individual opamps are high-performance, radiation tolerant devices (as would be an acceptable instrumentation amp). The radiation exposure effects of the resistor array are unknown. The resistors are the most critical component of the discrete structure.

Total Resistor Tolerance

Even precision resistors can't maintain the nominal tolerance; the effects of the accumulation of errors of a generic precision resistor factors which add to a total error budget are illustrated:

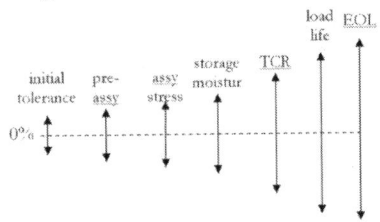

Figure 16: Generic Precision Resistor Cumulative Tolerance

Chapter 4: Numerical Example

Resistor Parameters

The resistors to be compared are the discrete and matched array of same manufacturing type, the Vishay Precision Group's "Type Z" thin-film resistors; both the discrete and array are military grade.

RNC90Z is discrete unit of RNC class; the PRND is a matched array of 4 resistors of the same material as the RNC90Z.

The worst-case performance of these resistors is given:

	RNC90Z	PRND
initial tolerance:	± 0.01 %	± 0.01 %
matching:	N/A	± 0.005 %
tempco:	± 2 ppm/°C	± 2.5 ppm/°C
load/life	± 0.05 % / 2000hrs 125°C	± 0.05 % / 2000hrs 70°C
matching	N/A	± 0.02 %
shelf life	± 25 ppm (tbl4-VI)	±0.01% (tbl1)

Table 1: Vendor (Vishay Precision Group) Resistor Specifications

Both resistors are from same manufacturer; both are of same material.

Aging Calculations:

The PRND does not have life figures for 10k hrs. Use 2000 hr values for equal comparison. (The RNC life figure for 10k hrs is 10x that for 2k hrs. Being of the same material, it may be feasible to assume the same for PRND ... but PRND is less robust; better to stick with stated values. Note that a 5x increase in time resulted in a 10x worsening of tolerance.)

PNRD: Flatpack matched array: 70 C at 150 mW/resistor

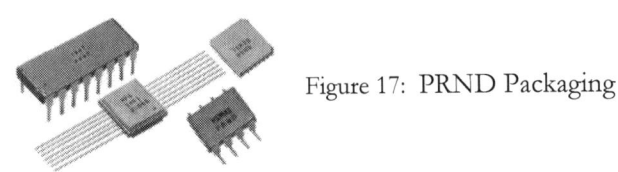

Figure 17: PRND Packaging

At 70°C: 0.05%/2k hrs

$$A_{70} = 0.05\% \text{ per 2k hr} \Rightarrow \frac{0.05}{100} \rightarrow 500 \text{ ppm}/2000 \text{ hr}$$
$$= 0.25 \text{ ppm/hr}$$

At 25°C: 2 ppm/yr (time length assumed)

$$A_{25} = 2 \text{ ppm/yr} \Rightarrow \frac{2 \text{ ppm}}{8760 \text{ hr}} \Rightarrow 228 \times 10^{-6} \text{ ppm/hr}$$

Per Arrhenius Equation ... where δ is the aging coefficient

$$\delta_{PRND} = \frac{A_{70}}{A_{25}}\left[\frac{10}{70-25}\right] \Rightarrow \delta = \frac{0.25}{228\times 10^{-6}}^{[0.222]} = 4.736$$

a far piece out from the rule-of-thumb of $\delta = 2$, eh?

And a calculation for the aging coefficient:

$$A_{60, PRND} = A_{25} \times \delta_{PRND} \wedge \left[\frac{T-T_o}{10}\right] \Rightarrow 2 \times 4.736 \wedge \left[\frac{60-25}{10}\right]$$

$$A_{60, PRND} = 0.0528 \text{ ppm/hr} \Rightarrow 462 \text{ ppm/yr}$$

RNC90Z: Discrete: 125iC at 300 mW

Figure 18: RNC90Z Packaging

At 125°C: 0.05%/2k hrs, 0.5%/10k hrs at 125°C, 300 mW

$$A_{125} = 0.05\% \text{ per 2k hr} \Rightarrow 500 \text{ ppm}/2000 \text{ hr}$$
$$= 0.25 \text{ ppm}/\text{hr}$$

At 25°C:

$$A_{25} = 25 \text{ ppm}/\text{yr} \Rightarrow 25 \text{ ppm}/8760 \text{ hr}$$
$$= 0.00285 \text{ ppm}/\text{hr}$$

$$\delta_{RNC} = \frac{A_{125}}{A_{25}}\left[\frac{10}{125-25}\right] \Rightarrow \delta = \frac{0.25}{2.854 \times 10^{-3}}^{[0.1]} = 2.013$$

The aging coefficient in ppm/yr:

$$A_{60,RNC} = (2.854 \times 10^{-3}) \times 2.013 \wedge \left[\frac{60-25}{10}\right]$$

$$\Rightarrow A_{60,RNC} = 52.8 \text{ ppm}/\text{yr}$$

Although it is not expected that the system will operate at this temperature until EOL, this is a worst-case analysis. Furthermore, the vendor specifications were tested under full-load.

Although the array has many "better" qualities than the discrete, it is also more susceptible to aging. This will become a significant factor in extreme environments. These calculations do not include radiation effects.

Worst-Case 7yr EOL Resistor Tolerance
RNC90Z Discrete

$$\%\Delta R_{RNC90} = it + \Delta T\ TC + \Delta Y\ a60rnc$$
$$= 0.01 + (60-25)\ 2.0 \times 10^{-4} + 7 \times 2.32 \times 10^{-3}$$

$$\%\Delta R_{tot} = +0.0332\%$$

PRNDMatched Array

$$\Delta R_{PRND} = it + \Delta T\ TC + \Delta Y\ a60prnd$$
$$= 0.01 + (60-25)\ 2.5 \times 10^{-4} + 7 \times 46.2 \times 10^{-3}$$

$$\Delta R_{tot} = +0.343\%$$

Although the PRND is a matched array, the construction is less robust than the RNC90. The tolerances are identical but the RNC has a 20% smaller temperature coefficient and more robust temperature range; only the unit-unit matching is better with the array.

However, when it comes to ratio matching, the PRND holds a ratio to 0.02%; the worst-case matching of the RNC90Z is 0.2%

Array Resistor Considerations

Under worst-case conditions, the matching is added to the error of the reference device. Array matching errors usually track together whereas several discrete devices track randomly. This is usually a benefit unless absolute values are important. A discrete device often has a tighter absolute tolerance.

Further analysis along these lines begins to delve into the manufacture of the components themselves; outside the scope of purpose. Such a topic would be of interest when designing integrated circuits.

Chapter 4: Numerical Example 43

Component Parameters

Opamp Nominal (max) Parameters (25 C)

$V_{os,typ}$ = 858V
I_{bp} = 2.5 nA
I_{bm} = -2.5 nA
I_{bos} = 2.5 nA

Resistor EOL (max) Tolerance (7 yr) (but matching ≤ 0.02%)

PRND = 0.342 % { 9967.63 (9965.6/9969.6), 10034.2 (10032.2/10036.2) }
RNC90Z = 0.033 % { 9998.18 , 10003.3 }

Offset Error

$$\varepsilon_{os} = v_{os} \frac{R4 + R6}{R6} \Rightarrow 170.0\,\mu V_{PRND}\,;\,170.6\,\mu V_{RNC}$$

Bias Current Error

$$\varepsilon_{ibp} = -i_{bp}(1-k)\frac{R5\,R7}{R6}\frac{R4+R6}{R5+R7}$$

$$\Rightarrow -25.10\,\mu V_{PRND}\,;\,-25.09\,\mu V_{RNC}$$

$$\varepsilon_{ibm} = i_{bm}\,R4 \Rightarrow -25.09\,\mu V_{PRND}\,;\,-25.09\,\mu V_{RNC}$$

Total Worst-Case Nominal DC Error

$$V_{o,DCerr} = -\frac{R6}{R4}\left[\varepsilon_{os} + |\varepsilon_{ibm} - \varepsilon_{ibp}|\right]$$

$$\Rightarrow -170.0\,\mu V_{PRND}\,;\,170.6\,\mu V_{RNC}$$

Apparently insignificant differences due to resistor type.
Note that bias current error is probably insignificant.

Gain & Common-Mode Rejection Ratio (CMRR):

The 1st-stage common-mode gain was defined as:

$$A_{CM1} = \frac{(R1 + R3)(A_{OL1} - A_{OL2})}{R1\,A_{OL2} + R2\,A_{OL1}\,A_{OL2} + R3\,A_{OL1}}$$

The magnitude of open-loop gain will be far greater than the resistor values. Furthermore, for all resistors being of equal value, the common-mode gain is a measure of open-loop gain mismatch. Even with resistor mismatch, the magnitudes of open-loop gain dominate.

$$A_{CM1} = \frac{(R + R)(A_{OL1} - A_{OL2})}{R\,A_{OL2} + R\,A_{OL1}\,A_{OL2} + R\,A_{OL1}}$$

$$= \left.\frac{2(A_{OL1} - A_{OL2})}{A_{OL2} + A_{OL1}\,A_{OL2} + A_{OL1}}\right|_{\text{Equal R}}$$

This may not be so in amplifiers with lower AOL and significant resistor mismatch

Given the assumed maximum open-loop gain mismatch of nominal and minimum, the common-mode gain of the 1st-stage is:

$$A_{CM1} = \frac{2(1.4 \times 10^9 - 300 \times 10^6)}{300 \times 10^6 + (1.4 \times 10^9)(300 \times 10^6) + 1.4 \times 10^9} = 5.238 \times 10^{-9}$$

The differential-mode gains of the two stages were defined as:

$$A_{DM1} = 1 + \frac{R1 + R3}{R2} \qquad A_{DM2} = -\frac{R6}{R4}$$

Resistor mismatch can be significant here.

Chapter 4: Numerical Example   45

Consider the effect of resistor tolerance on the 1st-stage differential gain. Assuming the resistors have equal nominal value, the possible perturbations due to tolerance vary as shown for k = 0 v 1.

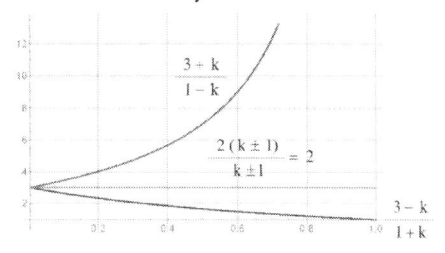

Figure 19: 1st-stage differential-gain variance

The differential-mode gain limits of the 2nd-stage are:

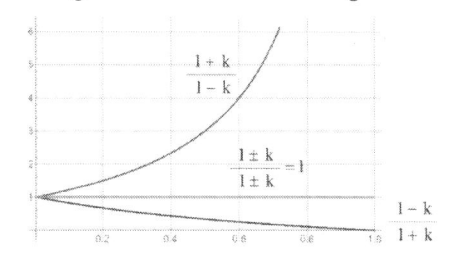

Figure 20: 2nd-stage differential-gain variance

The differential gain limits (for k < 5%) of the full amplifier network are:

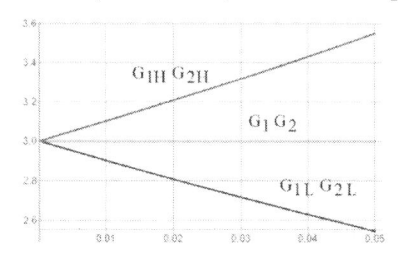

Figure 21: Full amplifier differential-mode gain variance

The 1st-stage common-mode gain is expressed:

$$A_{CM1} = \frac{(R1 + R3)(A_{OL1} - A_{OL2})}{R1\, A_{OL2} + R2\, A_{OL1}\, A_{OL2} + R3\, A_{OL1}}$$

For the A_{OL} extremes, A_{CM1} varies between 5.238E-9 and 5.240E-9 due to resistor mismatch of 0.02%

The 2nd-stage common-mode gain for resistors with identical tolerance limits:

$$A_{CM2wc} = \frac{-4k\, R6}{(1 - k)[R4(1 + k) + R6(1 - k)]}$$

However, if the nominal resistor values are equal, this reduces to:

$$A_{CM2wc} = -\frac{2k}{k - 1} \Rightarrow 400.08 \times 10^{-6} \bigg|_{k=0.02\%} = 67.96\, dB$$

For variations of resistor tolerance k from 0 to 1.0 (100%):

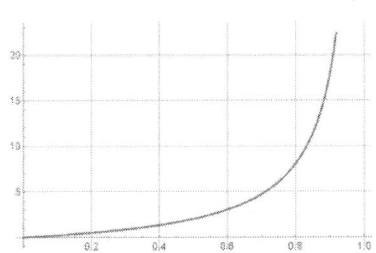

Figure 22: 2nd-stage common-mode gain variance due to resistor tolerance

The 2nd-stage common-mode gain consists of the sum of the imbalance due to resistor tolerance and that of the amplifier itself.

$$A_{CM2} = A_{CM2wc} + A_{CM2amp}$$

The minimum common-mode attenuation of the opamp is 120 dB:

$$120\, dB \equiv 10^{\wedge}-\frac{120}{20} \Rightarrow 10 \times 10^{-6}$$

Chapter 4: Numerical Example   47

The total 2nd-stage common-mode gain is therefore:

$$A_{CM2} = A_{CM2,res} + A_{CM2,amp}$$

$$= 400.08 \times 10^{-6} + 1.00 \times 10^{-6} = 401.08 \times 10^{-6} = -67.94 \text{dB}$$

It is clear the resistor mismatch dominates the 2nd-stage common-mode gain. The overall common-mode gain is found from:

$$A_{CM} = A_{CM1}\frac{R6}{R4} + A_{CM2}$$

$$A_{CM,T} = 5.24 \times 10^{-9} \times (1) + 401.08 \times 10^{-6} = 401.08 \times 10^{-6}$$

The excellent 1st-stage common-mode gain has no bearing on the overall system performance ...

The specifications for a representative integrated instrumentation amp are shown on the next page. In summary, the AD524C offers a minimum CMRR of 90 dB at gain of 3.

Representative Instrumentation Amplifier – AD524

A component-level instrumentation amplifier - the AD524S as an example - has a post-radiation minimum CMRR of 70 dB at a gain of 1. CMRR generally improves with gain; the AD524 $CMRR_{min}$ is 90 at gain of 10.

A gain of 3 as in the discrete network just examined is about mid-point of the log scale: CMRR ~ 80 dB at G = 3.

From the Analog Device's AD524 Data Sheet

Common-Mode Rejection DC to 60 Hz with 1kΩ Source Imbalance	CMR_{dB}
G = 1 ←———— G = 3 ————→	80
G = 10	100
G = 100	110
G = 1000	120

Figure 23: The log nature of the units suggest that
CMRR ~ 90 dB at G = 3
$log\, 3 = 0.477 \sim 0.5$

There are likely other parts with similar or improved performance specifications.

Use a manufactured part unless there's a good reason not to.
... and there's rarely a good reason not to.

Chapter 5: Phase Error

It is often feasible in precision circuits to adjust the initial DC offset errors to zero, but drift effects need to be considered. Specific parameters are dependent on the opamps selected. These procedures are estimates; any manufacturer's data will supersede these calculations.

Read the fine print in the data sheet.

The following have been determined:

Maximum tempco, RES: 2.5 ppm/°C_{PRND} ; 2 ppm/°C_{RNC}

Aging: 462 ppm/°C_{PRND} ; 120 ppm/°C_{RNC}

Maximum $V_{OStempco}$: 1 µV/°C	→	35 µV
Post-rad offset (300 krad 25°C):	→	110 µV
Maximum I_b tempco: 5 pA/°C	→	175 pA
Post-rad bias (300krad 25°C):	→	15 nA
Post-rad bias offset (300krad 25°C):	→	10 nA

The data sheet for this opamp does not provide matching data between units. In the absence of additional information, it will be assumed that this unit consists of 4 amplifiers on a common die rather than 4 individual units in a common package. As such, it will be further assumed that offset voltages and bias currents track. The worst-case conditions will be represented by offset mismatch rather than absolute mismatches.

Assume initial offset is pre-calibrated to zero and that tolerance variations are linear. This is usually valid ... but if always so, thermistors wouldn't need curvature compensation.

The worst-case offset voltage drift is given as 1 μV/°C with the nominal drift given as 0.1 μV/°C. With a maximum change in temperature ΔT of $(60 - 25) = 35°C$, the total differential offset drift will be 31.5 μV.

The offset voltages of two amplifiers vary over temperature change as:

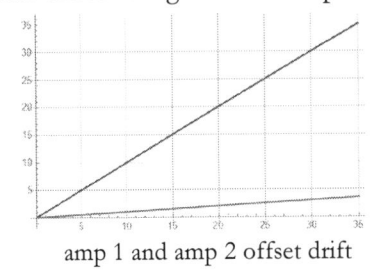

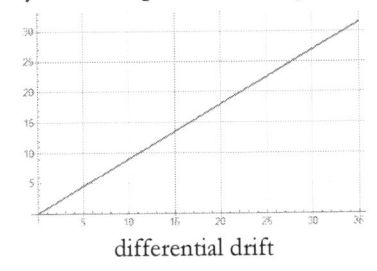

amp 1 and amp 2 offset drift   differential drift

Similarly for the bias currents:

The worst-case bias current drift is given as 3pA/°C with the nominal drift given as 0.42pA/°C. The total differential bias drift over temperature change will be 90.3pA.

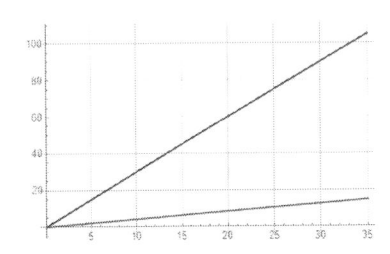

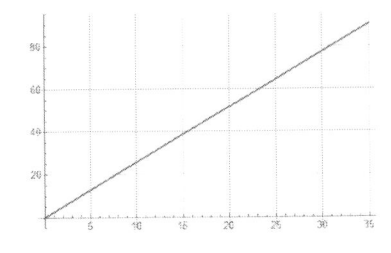

amp 1 and amp 2 bias offset drift   differential bias drift

The worst-case bias current variation is estimated as:

$$\frac{ib_{max}}{ib_{nom}} = \frac{5\,nA}{80\,pA} = 62.5$$

Assuming bias currents track with temperature, only the bias offset will be a factor, therefore:

$$\frac{ib_{os,max}}{ib_{os,nom}} = \frac{3\,nA}{80\,pA} = 37.5$$

Chapter 5: Phase Error

Taking into account worst-case resistor tolerances, the maximum DC offset error – assuming initial offset is calibrated to zero – is found as follows:

The 1st-stage offset error is derived from a combination of offset drift mismatch and the bias current mismatch through worst-case value resistors. This may be expressed as:

$$V_{os1} = \Delta T \Delta v_{os} + \Delta i_{os}(R1+R3)(1+k)\left(1 + \frac{(R1+R3)(1+k)}{R2(1-k)}\right)$$

For the values presented:

$$V_{os1} = 35 \times 10^{-6} + 56.8 \times 10^{-9}(1 + 2.44) = 35.2\,\mu V$$

The effects of bias current are negligible. This may not be true if the gain of the system were greater than 3. For example, if the 1st-stage gain were 100 (40 dB) using R1 = R3 = 100k and R2 = 1k, the offset would be 174 µV; the majority of which would be due to bias current offset.

The 2nd-stage offset is determined from:

$$V_{os2} = -\frac{R6}{R4}\left[vos'\,\frac{R4+R6}{R6} + \left|ibm\,R4 - ibp\,\frac{R5\,R7(R4+R6)}{R6(R5+R7)}\right|\right]$$

Again, for the values presented:

$$V_{os2} = -1.22\left[57.3 \times 10^{-6} + \left|945 \times 10^{-9} - 132 \times 10^{-9}\right|\right] = 71.0\,\mu V$$

The effects of bias current in the 2nd-stage are also negligible. The offset error at the output is found as:

$$V_{os} = V_{os2} + V_{os1}\,\frac{R6}{R4} = 114\,\mu V$$

The expression for phase error is derived from:

$$V(t) = V_p \sin(\omega t) \Rightarrow t = \frac{1}{\omega} \sin^{-1} \frac{V(t)}{V_p}$$

The time error related to offset is found to be:

$$t = \frac{\phi}{2\pi f} \Rightarrow \phi = \frac{360}{2\pi} \sin^{-1} \frac{V_{os}}{V_p}$$

Given: $V_{os\,max} = 114\mu V$, an overall amplifier gain of 3, and input voltage range of $3 \rightarrow 11$ mV.

$$\phi_{vin,min} = \frac{360}{2\pi} \sin^{-1} \frac{114 \times 10^{-6}}{3 \times 3 \times 10^{-3}} \Rightarrow 0.73°$$

$$\phi_{vin,max} = \frac{360}{2\pi} \sin^{-1} \frac{114 \times 10^{-6}}{3 \times 3 \times 10^{-3}} \Rightarrow 0.20°$$

While these phase shifts could usually be considered negligible, some scientific instruments seek to measure such phase differences.

Recall that these figures represent drift: it was assumed initial offset errors were calibrated to zero.

The phase drift error due to offset variations in this specific network may be minimal depending on the application, signal bandwidth, and input voltage range.

This being basically a DC analysis, care should be taken to consider frequency effects in a practical application. This would be particularly important if a filter were part of the signal chain.

Chapter 6: Supply Voltage regulation

Power-Supply Rejection Ratio (PSRR)

One issue not analyzed is the effect of power supply variations on the signal wherein fluctuations in the supply voltage may also have a detrimental effect on the performance of a high-precision amplifier system. The connection between power and signal is not readily apparent; it is a factor of the internal topology of the amplifiers.

The effect on a network may be estimated by examining the PSRR response provided in the amplifier data sheet. PSRR is similar to CMRR; PSRR is a measure of the ability to reject power supply variations. For example, PSRR of 80 dB implies that for a supply voltage variation of 100 mV, there will be a corresponding 10 µV shift in signal.

For comparison, the PSRR charts for the opamp of this network, the Intersil ISL70417, is compared to a discrete instrumentation amp, the Analog Devices AD524. Note that both provide PSRR data for both positive and negative supply voltages – amplifier topologies are often not "balanced" with respect to power connections. Also note that performance improves as gain increases.

Like both open-loop gain and common-mode rejection, the rejection ability decreases with frequency. A PSRR of 0 dB implies that a 1V change in power will cause a 1 V change in signal.

The information provided by Analog Devices shows typical PSRR for a variety of gains for a fixed supply and signal voltage. The least rejection occurs at gain of 1. Although Intersil only provides information for a gain of 1, it is reasonable to assume this represents the minimal performance.

PSRR of AD524

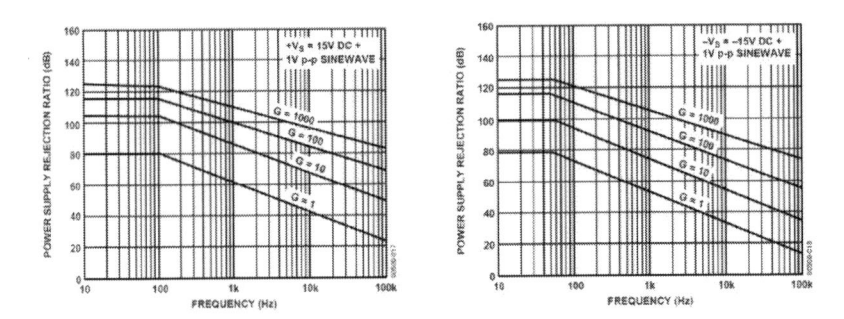

Figure 24: PSRR of AD524: To 60 Hz, the PSRR of both 15V supplies is 80 dB

PSRR of ISL70417

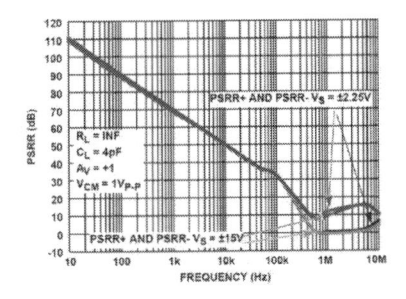

Figure 25: PSRR of ISL70417: At 10Hz, PSRR of a 15V supply is 110 dB, decreasing to 90 dB at 80 Hz.

The ISL70417 offers superior PSRR performance when compared to the AD524 ... but both have poor PSRR at higher frequencies.

Minimizing perturbations on the power supply lines is an even more effective means of reducing supply interaction with the signal. This is usually accomplished with power supply regulation.

For precision applications, linear regulators are preferred.

Chapter 6: Supply Voltage Regulation 55

LM117 Adjustable Positive Voltage Regulator

Consider a common power regulator network. I personally used this circuit for spacecraft applications in 1985; I have most recently seen the identical circuit used at NASA in 2012. A robust and well-characterized linear regulator network suitable for high-reliability in the extreme environment of space.

Linear regulators should be used in such applications to prevent clock-feedthrough possible with switching regulators. A switching regulator should be utilized as an off-board pre-regulator with a final linear regulator on-board with the precision network.

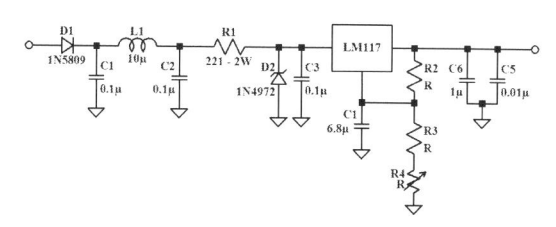

Figure 26: Power Supply LM117 Regulator Network

A side note regarding a bit of integrated analog development history:

Among a few others, Bob Dobkin, Bob Widlar, Bob Pease, and Jim Williams are among the very top tier of this specialty.

Robert Dobkin of National Semiconductor developed the LM117 in 1976. Mr. Dobkin is of the "old school" - he studied EE at MIT, but didn't complete his degree. The field was new — he learned by doing. Worked at Philbrick in IC design with Bob Pease. Ended up at National in 1969 where he became Director of Advanced Circuit Development. Left National Semi with Bob Widlar (inventor of the bandgap regulator) in 1981 to form Linear Technology where he is now the Chief Technical Officer — Jim Williams' former boss.

Bob Widlar died in 1991, Jim Williams died in 2011, Bob Pease died a week later returning from Jim Williams' memorial service.

Linear Technology remains ~~the~~ a premier analog component supplier to this day; National Semiconductor became part of Texas Instruments in 2011. Analog Devices is not a Silicon Valley phenomenon; they stayed on the east coast ... and are also of the top tier.

The basic operating network of the LM117 is shown:

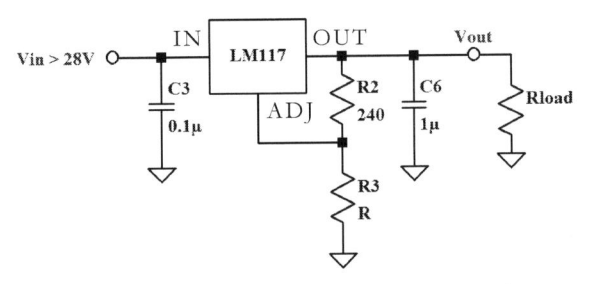

Figure 27: Basic LM117 Voltage Regulator

The output voltage is defined by:

$$V_{out} = 1.25 \left(1 + \frac{R3}{R2}\right) + I_{adj} R3$$

The device supplies 1.5A over an output voltage of 1.25 – 37V and temperature range of -55 – 150 TC. It offers short-circuit protection as well as current limiting, thermal overload, and SOA protection. Operation without capacitors is possible but the output capacitor offers improved transient response. The input capacitor can be neglected in the regulator is less than 6" from the power supply filtering caps.

Applying a capacitor (not shown) across R3 will significantly improve ripple rejection.

The regulator "floats" and may be operated at high common-mode voltages if the input-output voltage is not exceeded. High common-mode voltage operation requires the output not be shorted to GND.

Connecting the ADJ pin to ground (shorting R3) will force the output to 1.25V. Internal overload protection circuitry continues to function in this mode.

Chapter 6: Supply Voltage Regulation

Spacecraft power supplies are often specified at a nominal 28V with variation of ±6V. Many precision analog electronic networks are required to operate at 15V (or lower – perhaps at 5V?[2]). This places a maximum voltage drop of (28 + 6 – 15 = 19V) across the device – well within the specified input-output differential voltage of 40V. The minimum allowable voltage drop across the regulator is 3V. The minimum supply voltage is 22V leaving 7V headroom.

Assume a nominal output load current of 10.5 mA at 15V output. This implies a load resistance of:

$$R_{load,nom} = \frac{15 \text{ V}}{10.5 \text{ mA}} \Rightarrow \sim 1430 \ \Omega$$

Further assume a 10% increase as maximum worst-case load.

$$I_{out,max} = 1.1 \times 10.5 \text{ mA} \Rightarrow 11.6 \text{ mA}$$

$$R_{load,min} = \sim 1300 \ \Omega$$

... and a maximum regulator dissipation of:

$$P_{LM117,nom\,max} = 11.5 \text{ mA} \times 19 \text{ V} \Rightarrow 220.4 \text{ mW}$$

The regulator is to be operated well within its safe-area of operation.

The full regulator circuit to be analyzed is shown:

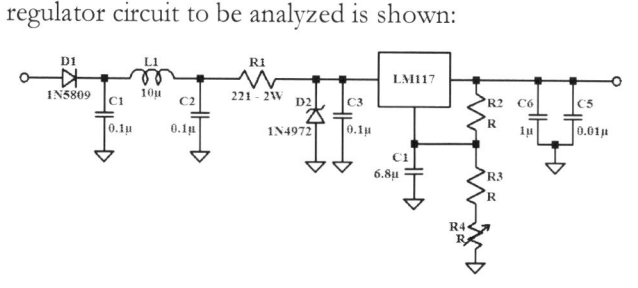

Figure 28: Power Supply Regulator Network

---

[2] One way of providing a split supply from a single-ended 5V rail is to level-shift the reference to VCC/2. The effective voltage supplies will be ±2.5V. One must be careful about "grounds" at other points of the circuitry though.

The DC equivalent regulator input circuit is:

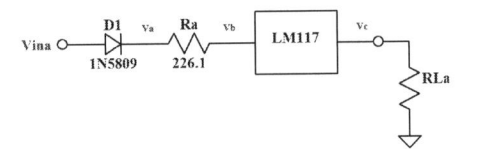

Figure 29: Regulator Equivalent Input Network

The worst-case forward voltage drop of the 1N5809 diode occurs at low temperatures. Extrapolating from the data sheet at 100 mA load, the maximum voltage drop will be is assumed to have a voltage drop of 0.70V at -15°C. At 60°C, the voltage drop will be 0.48V.

Resistor R1 is a 221 R, 1%, 2W device with a worst-case EOL tolerance of 1.8%. The maximum value of R1 will therefore be:

$$R1_{max} = 221(1 + 0.018) \Rightarrow 225\,\Omega$$

The 10 TH inductor has a maximum series resistance of 1.1 H. Equivalent series resistor Ra is the sum of the two resistances:

$$Ra_{max} = R1_{max} + RL_{max} \Rightarrow 225 + 1.1 \Rightarrow 226.1\,\Omega$$
$$Ra_{min} = R1_{min} + RL_{min} \Rightarrow 217 + 0.9 \Rightarrow 217.9\,\Omega$$

The worst-case condition for drop-out voltage will be for the output to be at maximum and input to be at minimum.

The LM117 has the following specifications:

Line regulation$_{max}$:	0.02%/V
Load regulation$_{max}$:	0.1% (with ΔI_L being negligible)
Temperature stability$_{max}$:	0.2% (-15°C → 60°C)
Life stability$_{max}$:	0.5%

It is assumed the output may be trimmed to 0.5%.

Empirical radiation testing at 10krad indicates negligible effect.

Chapter 6: Supply Voltage Regulation

The total tolerance of the 15V output can now be calculated as:

$$\Delta\%, V_{out} = (0.02 \times 6V) + 0.10 + 0.20 + 0.50 + 0.50 \Rightarrow 1.42\%$$

The maximum output voltage will be determined as:

$$V_{out,max} = 15(1 + 0.0142) \Rightarrow 15.21\,V$$

The minimum voltage allowed at the regulator input (Vb) is:

$$V_{b,min} = V_{out,max} + 3.0 \Rightarrow 18.21\,V$$

The minimum voltage allowed at node Va is:

$$V_{a,min} = V_{b,min} + I_{in}R_{a,max}$$
$$\Rightarrow 18.21 + (11.6\,mA)(226.1) = 20.83\,V$$

The minimum allowable input voltage is therefore:

$$V_{a,min} = V_{a,min} + V_{D1,max}$$
$$\Rightarrow 20.83 + 0.9 = 21.7\,V$$

The specification on minimum input voltage is 22V; the regulator should perform as desired under worst-case conditions. Because the regulator is operated under a very light load, the specified 3V input-output voltage differential is likely high; there is perhaps an unspecified safety margin input voltage margin of 0.5 → 0.75V but this is not guaranteed by analysis.

The maximum input voltage will occur under no load; ie, no input series resistance losses. Therefore, the maximum input voltage at the regulator input, Vb, will be:

The regulator design process has the goals of maintaining regulation under voltage and temperature variations and minimizing network power consumption. The required terminal voltages and currents have been defined but not the resistor values.

The regulator output voltage is defined by the following expression:

$$V_{out} = 1.25\left(1 + \frac{Ra}{R2}\right) + I_{adj} Ra$$

The "1.25" factor is the internal regulator reference voltage and Iadj is the effective regulator bias current out of the ADJ pin. Resistor Ra is a series combination of fixed R3 with a variable "lab-set" resistor R4.

The reference voltage has a tolerance of ∓50 mV; the bias current has a typical value of 50 0A. RNC90-type resistors will be used.

The maximum power dissipation will occur for maximum input voltage and load current and minimum diode drop, input series resistance, and output voltage.

To determine the value of Ra:

$$Ra\left(\frac{V_{ref}}{R2} + I_{adj}\right) = V_{out} - V_{ref}$$

$$\Rightarrow Ra = \frac{R2(V_{out} - V_{ref})}{V_{ref} + I_{adj} R2}$$

where $V_{ref} = 1.25 \pm 0.05$ and $I_{adj} = 50\ \mu A_{typ}$

Assume $R2 = 270\Omega \pm 0.1\%$ and a 15V output voltage.

The maximum acceptable value of Ra is found from:

$$Ra_{max} = \frac{270(1 + 0.001)(15 - 1.25 - 0.05)}{1.25 - 0.05 + 50 \times 10^{-6} \times 270(1 + 0.001)} = 3073.49\ \Omega$$

The minimum acceptable value is found to be:

$$Ra_{min} = \frac{270(1 - 0.001)(15 - 1.25 + 0.05)}{1.25 + 0.05 + 50 \times 10^{-6} \times 270(1 - 0.001)} = 2813.35\ \Omega$$

Chapter 6: Supply Voltage Regulation

Select a value for R3 such that R4 will have a suitable range of values for trim adjustments. A value of 9/10 of Ra is a reasonable starting point.

Set R3 = $0.9 \times Ra_{nom}$ = 2644.44 → R3 = 2610 Ω

(a 0.1%, 10 ppm/°C device of standard value is selected)

For a precise output, the value of R4 is variable where:

$R4_{range}$ = Ra_{max} − $R3_{min}$ → 3073.5 − 2610 = 463.5 Ω

The nearest standard value is 464 Ω, also a 0.1%, 10 ppm/°C device.

Calculating worst-case conditions:

Line reg$_{max}$:	0.02%/V x 6V	→	0.12 %
Load reg$_{max}$:	0.1%	→	0.10 %
Temp stability$_{max}$:	0.2%	→	0.2 %
Life stability$_{max}$:	0.5%	→	0.5 %
Total			1.42 %

Therefore, Vout = $15(1 \pm 0.0142)$ = {14.787, 15.00, 15.213}

With load current I_{load} = 11.6 mA, the power dissipation of the LM117 will be less than:

P max = 0.0115 (33.5 − 14.7) = 216.2 mW

This is well within the power rating of a TO-39 package.

The TO-39 package:

$$T_{j,max} = 150\,°C;\quad \theta_{ja} = 150\,°C/W;\quad \theta_{jc} = 15\,°C/W$$

With no heat sink, the LM117 max temperature will be:

$$T\,max = 150\,°C/W \times 0.216\,W = 32.4\,°C$$

Assuming no heat sink, the maximum ambient temperature will be 60°C.

The LM117 junction temperature will be:

$$T_{j,max} = T_{a,max} + T_{ja} = 60 + 32.4 = 92.4\,°C < 150\,°C$$

The maximum output voltage excursion will be 14.2 mV; power supply rejection of 80 dB will couple a power perturbation into the signal of:

$$80\,dB \equiv 10^{\wedge}-\frac{80}{20} \Rightarrow 10 \times 10^{-4}$$

$$100 \times 10^{-6} \times \Delta V_{supply} \Rightarrow \pm 1.42\,\mu V$$

This may not be an insignificant error depending on signal levels, the error tolerance of the measure, and the injection point in the signal chain.

Appendix A: Physical Connection

Note: An instrumentation amplifier will often be processing nV – μV level signals. The recommended input connection should be twin-ax shielded-twisted pair with active shielding per:

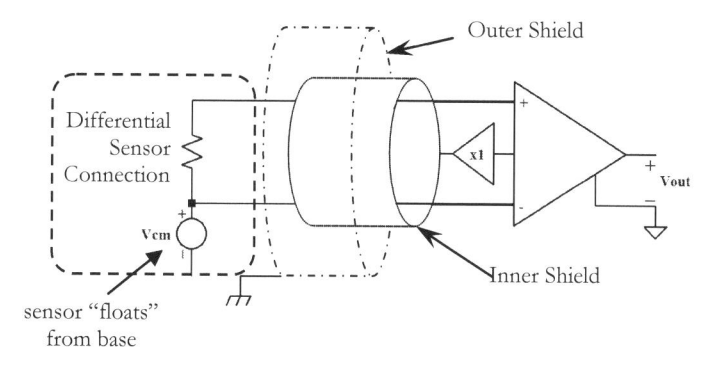

Figure 30: Suggested External IA Connections

Twin-ax cable with two isolated shields. Inner shield is actively driven by unity-gain buffer amplifier connected at common-mode point of IA – typically a mid-point of gain setting resistor R2 of the earlier example. This shield is unconnected at transducer end.

Transducer is electrically isolated from chassis and system ground. The chassis ground is connected to the outer shield which places a capacitively-coupled common-mode voltage between the shields. This outer shield is unconnected at the amplifier end.

The transducer signal appears as a differential signal; the potential differences between shields effectively forces noise and other interference to appear as a common-mode voltage.

The inner shield is driven by the common-mode voltage. The differential signal is isolated from chassis ground at either end and any noise coupled onto the inner shield is manifested as common-mode signal. The IA CMRR will reduce this effect from the measured signal.

Variations may exist depending on specific sensor construction.

Appendix B: Mathematica Snippets

Some example Mathematica snippets that were used in this development:

Define Parameters

Component Values

```
Clear[r1, r2, r3, r4, r5, r6, r7];
r1 = 10. × 10³;
r2 = 10. × 10³;
r3 = 10. × 10³;
r4 = 10. × 10³;
r5 = 10. × 10³;
r6 = 10. × 10³;
r7 = 10. × 10³;
rr = 10. × 10³;
```

Environment

```
dY = 7.;
dThi = 35.;
dTlo = -40.;
```

Amplifier DC Parameters:
Open-Loop Gain, Offset Voltage, Bias Currents

ISL70417SEH Quad package

```
vos = 10. × 10⁻⁶;  (* typical *)
vosmax = 75. × 10⁻⁶;  (* without radiation *)
vosmaxR = 110. × 10⁻⁶;  (* with radiation *)
ibmax = 5. × 10⁻⁹;
ibtyp = 0.08 × 10⁻⁹;
ibostyp = 0.08 × 10⁻⁹;
ibosmax = 3. × 10⁻⁹;
ibtc = 5. × 10⁻¹²;  (* per degC *)
cmrrAtyp = 145.;
cmrrAmin = 120.;
gbwp = 1.5 × 10⁶;
aOLAmin = 300 000 000.;
aOLAtyp = 1 400 000 000.;
aOLAdb = 20 Log[10, Abs[%]];
aOLBdb = aOLAdb + 10.;
aOLBmin = 10^(%/20);

ibm = ibtyp;
ibp = ibosmax;

kr = totRNChi;
kp = totPRNDhi;
km = 0.00025;
```

Appendix B: Mathematica Snippets

Resistor Aging Parameters - Need figures in ppm/hr

```
loadRNCpc = 0.05; (* per 2000 hrs at 125 C *)
shlfRNCpc = 0.0025; (* at 175 C *)
loadRNCppm = loadRNCpc 10^4;
shlfRNCppm = shlfRNCpc 10^4;

loadPRNDpc = 0.05; (* per 2000 hrs at 70 C *)
shlfPRNDppm = 2.0; (* at 70 C *)
loadPRNDppm = loadPRNDpc 10^4;
```

RNC Aging

$$a125rnc = \frac{loadRNCppm}{2000.};$$

$$a25rnc = \frac{shlfRNCppm}{8760};$$

$$\delta rnc = \left(\frac{a125rnc}{a25rnc}\right)^{\frac{10}{125-25}};$$

$$a60rnc = a25rnc \, \delta rnc^{\frac{60-25}{10}} \, 8760; \quad (* \text{ ppm/yr } *)$$

```
itrnc = 0.01; (* initial tolerance in percent *)
tcrnc = 2.0×10^-4; (* temperature coefficient in %/C *)
ycrnc = a60rnc 10^-4; (* convert aging coefficient from ppm/hr to %/yr *)
totRNChi = (itrnc + (dThi tcrnc) + (dY ycrnc))
(* EOL total hi tolerance at high temp in percent*)
totRNClo = -(itrnc + (dTlo tcrnc) + (dY ycrnc));
(* EOL total lo tolerance at low temp in percent*)
```

0.100736

PRND Aging

$$a70prnd = \frac{loadPRNDppm}{2000};$$

$$a25prnd = \frac{shlfPRNDppm}{8760};$$

$$\delta prnd = \left(\frac{a70prnd}{a25prnd}\right)^{\frac{10}{70-25}};$$

$$a60prnd = a25prnd \, \delta prnd^{\frac{60-25}{10}} \, 8760; \quad (* \text{ ppm/yr } *)$$

```
itprnd = 0.01; (* initial tolerance *)
tcprnd = 100.×2.5×10^-6; (* temperature coefficient in percent *)
ycprnd = 100 a60prnd 10^-6; (* aging coefficient in percent per year*)
ycmprnd = 0.015; (* match aging coefficient in percent per 2000hrs *)
totPRNDhi = (itprnd + (dThi tcprnd) + (dY ycprnd))
(* EOL total hi tolerance at high temp in percent*)
totPRNDlo = -(itprnd + (dTlo tcprnd) + (dY ycprnd));
(* EOL total lo tolerance at highlow temp in percent*)
```

0.342431

Equations

```
(*k=totRNChi;*)
(*k=totPRNDhi;*)
k = km;
```

$$aDM1 = 1 + \frac{(r1 + r3)}{r2};$$

$$aCM1 = \frac{(aOLAtyp - aOLAmin)(r1 + r3)}{aOLAmin\ r1 + aOLAtyp\ aOLAmin\ r2 + aOLAtyp\ r3};$$

$$aDM2 = -\frac{r6}{r4};$$

$$aCM2a = \frac{r4\ r7 - r5\ r6}{r4\ (r5 + r7)};$$

$$aCM2 = \frac{-4\ k\ r6}{(1-k)(r4 + k\ r4 + r6 - k\ r6)} + 10^{-\frac{cmrrAmin}{20}};$$

```
aDM = aDM1 aDM2;
Print["The differential gain of the network is ", Abs[aDM]]
```

$$\text{EngineeringForm}\left[aCM = aCM1\ \frac{r6}{r4} + aCM2\right];$$

```
Print["The common-mode gain of the network is ", EngineeringForm[aCM]]
```

$$cmrr = \frac{aDM}{aCM};$$

```
20 Log[10, %];
Print["The CMRR of the network is ", 20 Log[10, cmrr], " dB"]
```

$$\text{EngineeringForm}\left[voDCerr = -\frac{r6}{r4}\left(vos\ \frac{r4+r6}{r4} + \text{Abs}\left[ibm\ r4 - ibp\ \frac{r5\ r7}{r6}\ \frac{r4+r6}{r5+r7}\right]\right)\right];$$

```
Print["The untrimmed offset voltage at the output is ", EngineeringForm[voDCerr 10^6], " μV"]
```

```
The differential gain of the network is 3.
The common-mode gain of the network is -499.12×10^-6
The CMRR of the network is 75.5783 dB
The untrimmed offset voltage at the output is -49.2 μV
```

Define Plots

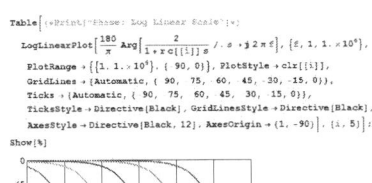

```
c = {100. ×10^-12, 1000. ×10^-12, 10000. ×10^-12, 0.1×10^-6, 1. ×10^-6};
clr = {Red, Green, Blue, Cyan, Magenta};
```

$$\text{Table}\left[(*Print["Magnitude: Log Scale"]*)\right.$$
$$\text{LogLinearPlot}\left[\left\{\frac{3}{\sqrt{2}},\ \text{Abs}\left[1 + \frac{2}{1 + r\ c[[i]]\ s}\right]\right\} /.\ s \to j\ 2\ \pi\ f\right],$$
$$\{f, 1, 1. \times 10^6\},\ \text{PlotRange} \to \{\{1, 1. \times 10^6\}, \{0, 4\}\},$$
$$\text{PlotStyle} \to \{\text{Directive}[\text{Black, Dashed}], clr[[i]]\},$$
$$\text{GridLines} \to \text{Automatic, GridLinesStyle} \to \text{Directive}[\text{Black}],$$
$$\left.\text{AxesStyle} \to \text{Directive}[\text{Black, 12}]\right],\ \{i, 5\}\right];$$
```
Show[%]
```

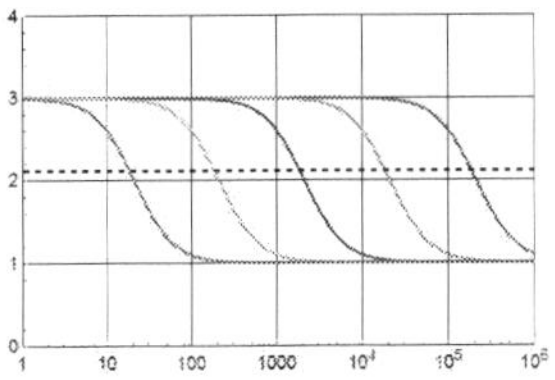

$$\text{Table}\left[(*Print["Phase: Log Linear Scale"]*)\right.$$
$$\text{LogLinearPlot}\left[\frac{180}{\pi}\ \text{Arg}\left[\frac{2}{1 + r\ c[[i]]\ s}\right] /.\ s \to j\ 2\ \pi\ f\right],\ \{f, 1, 1. \times 10^6\},$$
$$\text{PlotRange} \to \{\{1, 1. \times 10^6\}, \{-90, 0\}\},\ \text{PlotStyle} \to clr[[i]],$$
$$\text{GridLines} \to \{\text{Automatic}, \{-90, -75, -60, -45, -30, -15, 0\}\},$$
$$\text{Ticks} \to \{\text{Automatic}, \{-90, -75, -60, -45, -30, -15, 0\}\},$$
$$\text{TicksStyle} \to \text{Directive}[\text{Black}],\ \text{GridLinesStyle} \to \text{Directive}[\text{Black}],$$
$$\left.\text{AxesStyle} \to \text{Directive}[\text{Black, 12}],\ \text{AxesOrigin} \to \{1, -90\}\right],\ \{i, 5\}\right];$$
```
Show[%]
```

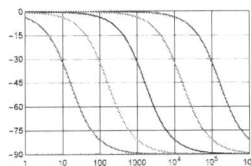

Appendix C: Opamp Fundamentals

The ideal opamp has the following properties:
- infinite open-loop gain
- infinite input impedance
- zero output impedance
- infinite bandwidth
- Vout = 0 for Vin = 0

Although practical opamps do not have these characteristics, for the most part, it is the feedback network which defines the network performance.

An opamp may be connected in one of two configurations depending on which input terminal the input signal is applied. The feedback connection of both is such that the ideal voltage difference between the amplifier input terminals is zero and that the potential at the non-inverting input (the "+" pin) is reflected to the inverting input (the "–" pin).

Inverting Configuration

With infinite input impedance, there is no current flowing into the amplifier terminals.

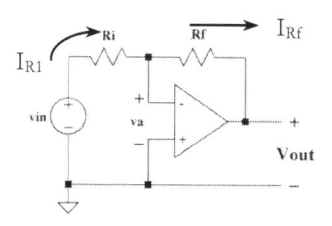

With infinite gain, the network expressions become:

$$I_{Ri} = I_{Rf} \Rightarrow \frac{v_{in} - v_a}{R_i} = \frac{v_a - v_{out}}{R_f} \Rightarrow V_{out} = -\frac{R_f}{R_i}$$

where $A_{OL} = \infty$ $V_{out} = -v_a A_{OL}$

and $v_a = -\dfrac{v_{out}}{A_{OL}} \rightarrow 0$ for $A_{OL} \rightarrow \infty$

Non-Inverting Configuration

Feedback is still applied to the inverting node; the potential at this node remains that of the + input (where va = 0). The resistor current directions are reversed but I_{Rf} is still equal to I_{Ri}.

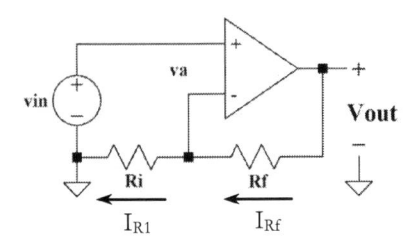

Again with infinite gain, the network expressions become:

$$I_{Rf} = I_{Ri} \Rightarrow \frac{v_{out} - v_a}{Rf} = \frac{v_a}{Ri} \Rightarrow V_{out} = 1 + \frac{Rf}{Ri}$$

where the same assumptions as in the inverting case hold true.

Note that the gain of this topology cannot be less than 1 ... which occurs when Rf → 0 and Ri → ∞ as shown:

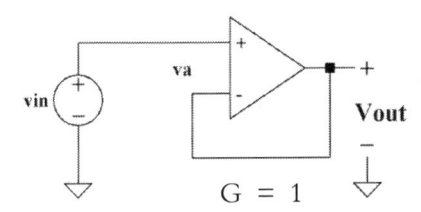

Any voltage appearing on the +input will be amplified by at least a factor of 1

Appendix C: Opamp Fundamentals

Open-Loop Gain A_{OL}

While modern opamps may have very large open loop gains – in some cases upwards of 140 dB (G = 10,000,000), it's not infinite. As desired signal resolutions become smaller, the assumption that $A_{OL} \rightarrow \infty$ becomes strained.

Define a feedback factor β such that:

$$\beta = \frac{R1}{R1 + Rf}$$

The expressions for the gains of the two opamp configurations become:

For the inverting case:

$$G_{inv} = -\frac{Rf}{R1} \cdot \frac{1}{1 + \frac{1}{A\beta}} = \frac{1 - \frac{1}{\beta}}{1 + \frac{1}{A\beta}} = \frac{A(\beta - 1)}{1 + A\beta}$$

where $A \equiv A_{OL}$ for convenience in notation.

For the non-inverting case:

$$G_{noninv} = \frac{1 + \frac{Rf}{Ri}}{1 + \frac{1}{A\beta}} = \frac{\frac{1}{\beta}}{1 + \frac{1}{A\beta}} = \frac{A}{1 + A\beta}$$

In both cases, the expressions for closed-loop gain may be stated as:

$$G = G_{ideal}\left(1 + \frac{1}{A\beta}\right)$$

where the gain error is represented by $1/(A\beta)$.

Factor β is defined by the external network components; factor A is inherent to the amplifier. Variations in A due to factors such as temperature place a limit on the minimum error.

This expression is worth a look at a practical example:

Consider an amplifier with closed-loop gain of 10. The selected opamp has an open-loop gain of 100 dB.

$$\text{Since} \quad G = 1 + \frac{Rf}{Ri} \Rightarrow 10 = 1 + \frac{Rf}{Ri} \Rightarrow \frac{Rf}{Ri} = 9$$

$$\beta = \frac{R1}{R1 + Rf} = \frac{1}{G} = 0.1$$

An open-loop gain of 100 dB is equivalent to a gain of 100,000.

The resulting DC gain error would be:

$$G = G_{ideal}\left(1 + \frac{1}{A\beta}\right) = 10\left(1 + \frac{1}{(100000)(0.1)}\right) = 10.001$$

$$\varepsilon_G = 0.01\%$$

This seems acceptable ...

Consider this opamp. It has open-loop gain of over 160dB ... even better than the above example - but over frequency, the open-loop gain decreases to 60 dB at 1 kHz.

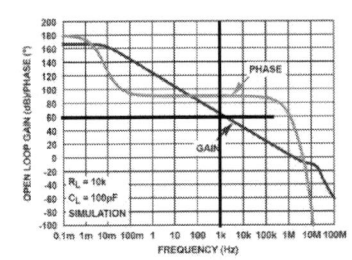

If this amplifier were to be used for a signal bandwidth of 1 kHz, the gain-error for signals at 1 kHz would be:

$$G_{1kHz} = 10\left(1 + \frac{1}{(1000)(0.1)}\right) = 10.1$$

Not quite as good ...

This is a 1% gain error; generally not considered suitable for precision measurements. This amplifier is most suited for low-frequency signals or lower closed-loop gain..

Appendix C: Opamp Fundamentals

Offset Voltage

By definition, the offset voltage of an opamp is that voltage applied to the non-inverting input which forces the output to zero volts with a zero-volt input. It is modeled as shown in the illustration below.

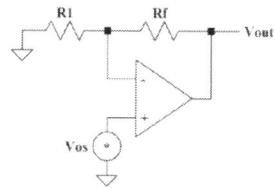

The offset voltage appears as a signal input for the non-inverting topology and is amplified as such.

This offset is generally due to imbalance and mismatching of elements of the opamp's internal network. The voltage may be of either polarity and typically ranges from low TV to mid mV.

Although either input is subject to offset voltages, it is usually easiest to model as a voltage source at the non-inverting input when mathematically; it represents an input voltage of an ideal non-inverting topology. The voltage at the output is determined by:

$$V_{out,os} = V_{os}\left(1 + \frac{Rf}{R1}\right)$$

If the amplifier gain is large enough, the offset voltage alone will cause the amplifier to become saturated.

Offset voltages are temperature dependent. It is usually assumed the offset voltage drifts linearly with temperature, although this is not necessarily true. The temperature coefficient is defined as:

$$V_{osTC} = \frac{\partial V_{os}}{\partial T} \Rightarrow \frac{\Delta V_{os}}{\Delta T}$$

The change in voltage may be predicted by: $V_{os}(T) = \Delta T \times V_{osTC}$

Bias Current

The amplifier bias currents may also create error. Although less of a problem with CMOS inputs than bipolar, the issue may still be significant.

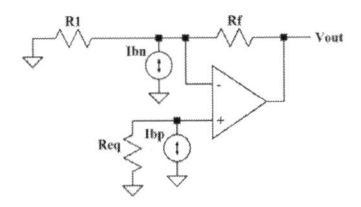

The ideal opamp analysis assumes no current flows into the amplifier inputs. Although the actual bias current is often small, it gets converted to a voltage by the network resistances and amplified with the signal.

Bias current may flow in either direction. It may often be assumed that the amplifier inputs are balanced which implies the bias currents are balanced as well, but bias current may also be offset from each other. Bias current magnitudes may range from fA to uA, depending on the amplifier characteristics.

The bias current offset is defined as: $I_{b,os} = I_b^+ - I_b^-$

The effect on the amplifier network may be determined by superposition:

For I_b^+: $\quad V_{out,ib} = I_b^+ R2 \left(1 + \dfrac{Rf}{R1}\right)$

This is identical to an input voltage of $R2 I_b^+$ at the input of a non-inverting amplifier.

For I_b^-: $\quad V_{out,ib} = - I_b^- Rf$

This is identical to a transimpedance amplifier: the output voltage is that of the input current I_b^+ passing through the feedback resistor R3.

Appendix C: Opamp Fundamentals

Note that the gains at each input are different. While it should be obvious that bias current effects are reduced by smaller resistor values, this is contrary to the goal of low power (from the larger currents for a given voltage drop).

Because the resulting voltage polarities are different, they tend to offset each other. To null the bias current error, the resulting voltages should have equal magnitudes.

If the nominal bias currents are assumed equal, then: $I_b^+ = I_b^+ = I_b$

so that: $I_b \, Rf \equiv I_b \, Req \left(1 + \dfrac{Rf}{R1}\right)$

Solving for Req: $Req = \dfrac{R1 \, Rf}{R1 + R1}$

Under ideal conditions, if R2 is set equal to the parallel combination of R1 and Rf, the effects of bias current error goes to zero.

$$V_{out,os} = I_b \, Rf - I_b \, Req \left(1 + \dfrac{Rf}{R1}\right)$$

$$= I_b \, Rf - I_b \dfrac{R1 \, Rf}{R1 + R1}\left(1 + \dfrac{Rf}{R1}\right)$$

$$\Rightarrow I_b \, Rf - I_b \, Rf = 0$$

It is not unusual for bias currents to be different from each other by 100%, although values of 10 – 20% are more typical. Furthermore, the bias currents and the offset current tend to drift with temperature. This is unpredictable and not easily corrected. The most straight-forward solution is to select an amplifier with low bias current specifications.

Common-Mode Gain

Keeping in mind that "ground" (GND) is simply a reference point from which other node voltages are measured, the GND nodes of the previous networks may actually be at an arbitrary potential. This potential is the common-mode voltage. In a differential network, this voltage may be defined as the average of the two input voltages:

$$vcm = \frac{V1 - V2}{2}$$

Although the ideal difference amplifier has zero common-mode gain, the actual common-mode voltage causes an output voltage of:

$$V_{out,cm} = vcm \, A_{cm}$$

This is often considered in terms of the common-mode rejection ratio (CMRR) in which:

$$CMRR = \frac{G}{A_{cm}}$$

where G is the gain of the closed-loop network; it may also be considered the differential-mode gain: $G \equiv A_{dm}$.

CMRR is often expressed in dB.

$$CMRR_{dB} = 20 \, log \, \frac{G}{A_{cm}}$$

Neither the differential or common mode gains are truly linear.

For the following network:

$$A_{cm} = \frac{1 - \frac{1}{CMRR}}{1 + \frac{1}{A}} = \frac{A \, (CMRR - 1)}{CMRR \, (A + 1)}$$

Common-mode effects are a significant cause of error.

Appendix C: Opamp Fundamentals

Input Impedance (Resistance)

The following model represents the significant elements of the input resistance:

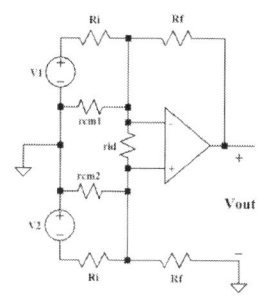

The resistance between the two inputs is the differential input resistance; the resistances from either input to the common terminal are the common-mode resistances (assumed equal).

The effect on the output is expressed:

$$V_{out} = V2 \frac{1 + \frac{Rf}{Ri} + \frac{Rf}{Rcm}}{1 + \frac{1}{A\beta'}} - V1 \frac{\frac{Rf}{Ri}}{1 + \frac{1}{A\beta'}}$$

$$= \frac{A\beta'}{Ri(1 + A\beta')} \left(V2 \frac{rcm\,R1 + R1\,Rf + rcm\,Rf}{rcm} - V1\,Rf \right)$$

where:

$$\beta' = \frac{1}{1 + \frac{Rf}{Ri} + \frac{Rf}{rid} + \frac{Rf}{rcm}}$$

A current now flows into the non-inverting input:

$$i_{in+} = V2 \left[\frac{1 + \frac{Rf}{Ri} + \frac{Rf}{Rcm}}{A\,R_{id}\left(1 + \frac{1}{A\beta'}\right)} + \frac{1}{R_{cm}} \right]$$

A current also flows into the inverting input:

$$i_{in-} = V1 \frac{1}{Ri} \left[1 - \frac{\frac{Rf}{Ri}}{A\left(1 + \frac{1}{A\beta'}\right)} \right]$$

This represents an input imbalance but for most applications, the effects of the finite input resistance are negligible at low frequencies. Although this discussion does not consider AC analysis, the effects of input capacitance at higher frequencies can be significant.

As an example, the Analog Devices ADA4897 is a military grade low-power opamp. The common-mode input resistance is 10M; the differential input resistance is 10k. The open-loop gain is 100 dB.

Consider a configuration in which Ri = Rf = 10k. V2 represents the common-mode voltage; V1 is the input voltage. If the common-mode voltage is 1V and the input signal is 2V (a 1V differential input), then:

$$i_{in+} = 204 \text{ nA} \qquad i_{in-} = 1 \text{ nA}$$

This could represent a substantial imbalance error in some networks.

Appendix C: Opamp Fundamentals

Output Impedance (Resistance)

Along with the other effects of reality vs. practical, the output resistance of an opamp is not zero. This may be modeled as:

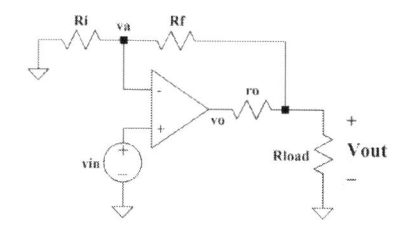

The network equations are modified as:

$$vo = A(vin - \beta vout) \quad \text{where} \quad \beta = \frac{R1}{Ri + Rf}$$

which leads to:

$$V_{out} = vo \frac{1}{1 + \frac{ro}{R_{load}} + \frac{ro}{Ri + Rf}}$$

$$\Rightarrow V_{out} = A'(vin - \beta V_{out})$$

where:

$$A' = A \frac{1}{1 + \frac{ro}{R_{load}} + \frac{ro}{Ri + Rf}}$$

which means the open-loop gain is reduced by a factor of:

$$\frac{1}{1 + \frac{ro}{R_{load}} + \frac{ro}{Ri + Rf}}$$

Consider $ro = 10\Omega$, $Ri = Rf = 10k$, and $R_{load} = 2k$

$$\frac{1}{1 + \frac{ro}{R_{load}} + \frac{ro}{Ri + Rf}} \Rightarrow \frac{1}{1 + \frac{10}{2000} + \frac{10}{20,000}} = 0.9945$$

A bit over 0.5 % gain error.

Noise

A topic worthy of a booklet all its own ...

The inherent noise of the amplifier may be represented with equivalent noise sources as shown:

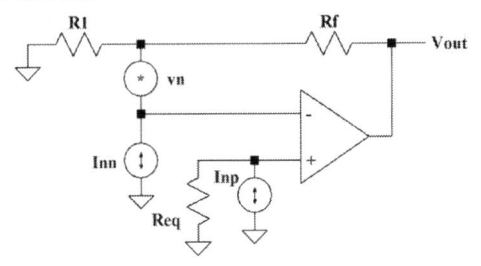

Given this network, the expression for noise as it appears at the output is:

$$vno_{RMS} = \left(1 + \frac{Rf}{R1}\right)\sqrt{Req\ inp^2 + vn^2 + Rf\ inn^2}$$

This expression assumes the noise sources have a normal (or Gaussian) probability distribution and that network bandwidth has been considered.

The Intersil ISL70417 has low-frequency noise of 0.25 TV; voltage noise density of 8.2 eV//Hz, and current noise density of 0.1 pA///Hz.

For the circuit above with bandwidth of 20 kHz and R1 = Rf = 10k

$$vno_{RMS} \Rightarrow 2.37\ \mu V$$

The peak-to-peak noise magnitude would be 14.2 µV.

The situation merits deeper consideration when thinking of quantization of very small signals...

One final note about data sheets: the first page belongs to Marketing; the inner pages belong to Engineering. While the information is (usually) correct on the front page, the data provided in tables within a section often called "Electrical Specifications" is more specific and should take higher priority in analysis.

References

Basically, the analysis within represents my own work ... in the sense that the homework I once did in Circuits class was also my own work.

I learned this material and techniques from any number of sources – as far back as the vacuum tube era Popular Electronics, various Heathkit instruction manuals, and Boy Scout merit badges. Any "analog" circuits text will present the basics; any good algebra text will provide the methodology. Ohm (1827), Kirchhoff (1845), and Thevenin (1883) pretty much covered it.

However ... there were instructors of note through the ages and one should never forget "on-the-job-training" where much of the "non-textbook" learning occurred. My deepest thanks to all of them.

> Paul Brunner, ham operator
> Robert Brooks, ham operator
> Paul Kruger, 10th-grade Electronics
> Allan Paul, EG&G, Las Vegas, NV
> Bruce Johnson, Robert Manhardt, University of Nevada-Reno
> Garry Cleveland, Lockheed Space Systems, Sunnyvale, CA.
> Lawrence Huelsman, University of Arizona

And a few textbooks in particular ... showing my age a bit:

> Kerr, RB, 1977, *"Electrical Network Science"*
> Hayt, W.H., Kemmerly, J.K., 1971, *"Engineering Circuit Analysis"*
> Graeme, J.G., Tobey, G.E., Huelsman, L.P., 1971,
>   *"Operational Amplifiers: Design and Applications"*

About the Author

Dave McGlone received BSEE and MSEE degrees focusing on analog signal processing techniques for low-level, high-precision sensor interface networks at discrete and integrated circuit levels. He obtained a PhD in electromagnetic geophysics from the Colorado School of Mines with a focus on geophysical instrumentation. He has over 30 years experience developing geophysical instrument systems including time spent with DoD and DoE ground and space-based projects. He recently retired from NASA's Langley Research Center where his primary work involved the development of photo-detection networks for laser-based measurement systems.

"Dave" sounds better than "Doctor"; Grandad was "Doctor" ...

Printed in Great Britain
by Amazon